ALLE ZEIT WACH
1842

Graphic on cover from Bellm, H.: "High-frequency characteristics of impulse-voltage measuring systems considering the transient radiation field", Ph.D. thesis, Faculty of El. Engineering University of Karlsruhe, 1981

Adolf J. Schwab

Field Theory Concepts

Electromagnetic Fields
Maxwell's Equations
grad, curl, div etc.

Finite-Element Method
Finite-Difference Method
Charge Simulation Method
Monte Carlo Method

With 49 Figures

Springer-Verlag Berlin Heidelberg NewYork
London Paris Tokyo

Prof. Dr.-Ing. Adolf Schwab
Institut für Elektroenergiesysteme und Hochspannungstechnik
Universität Karlsruhe
Kaiserstr. 12
7500 Karlsruhe 1

ISBN 3-540-18893-2 Springer-Verlag Berlin Heidelberg NewYork
ISBN 0-387-18893-2 Springer-Verlag NewYork Heidelberg Berlin

Library of Congress Cataloging-in-Publication Data
Schwab, Adolf J.
[Begriffswelt der Feldtheorie. English]
Field theory concepts : electromagnetic fields. Maxwell's equations, grad, curl, div, etc.
finite-element method, finite-difference method, charge simulation method, Monte Carlo method /
Adolf J. Schwab. -- [1st ed.]
Rev. translation of the 2nd German ed. of: Begriffswelt der Feldtheorie.
ISBN 0-387-18893-2 (U.S.)
1. Field theory (Physics) 2. Electromagnetic fields. 3. Maxwell equations. 4. Mathematical pyhsics.
I. Title.
QC173.7.S3813 1988
530.1'4--dc 19 88-24901

Offsetprinting: Color-Druck, Dorfi, GmbH, Berlin; Bookbinding: Lüderitz & Bauer, Berlin

2161/3020 543210 – Printed on acid-free paper

No mathematical purist could ever do the work involved in Maxwell's treatise. He might have all the mathematics, and much more, but it would be to no purpose, as he could not put it together without the physical guidance. This is in no way to his discredit, but only illustrates different ways of thought.

Oliver Heaviside

Preface to the English Edition

In translating his book Field Theory Concepts form the original German 2nd edition, published by Springer publishers in 1987, the author has added a number of revisions and late developments in the field. Thus, this translation represents essentially a revised edition of the German text, adapted for American readers. Due to the different historical development of concepts in the United States and in Europe, the author anticipates occasional discussions on his points of view.

The format of this book will likely seem unusual to many American readers. This is partially a product of different styles between the continents, and, in the author's opinion, is not altogether bad. More importantly, however, the new format represents a new approach to the teaching and understanding of electromagnetic field theory. An important example of this is the emphasis placed on the static, quasistatic, and dynamic nature of fields.

In trying to present the most lucid explanation of some difficult concepts, the author has tried to use highly illustrative terms. In some cases, this may necesitate an extra effort on the part of the reader to translate, mentally, his preconceived notions into the author's conceptual language. No apology is made for this, as the author is confident that this effort will be richly rewarded by a fuller understanding.

Regarding this Engish edition the author acknowledges his appreciation to Prof. Dr. Roland Schinzinger, University of California, Irvine, Prof. Dr. Chathan Cooke, Massachussets Institute of Technology, and Prof. Dr. Peter L. Levin Worcester Polytechnic Institute for numerous valuable suggestions and for thorough reading and polishing of the manuscript. Of course, any mistakes and omissions are ultimately my own.

Karlsruhe, June, 1988 Prof. Dr. Adolf J. Schwab

Preface

Maxwell's equations constitute the theoretical basis for the entire world of electrical and electronic engineering. Frequently, they appear only implicitly, for instance in control theory or in digital electronics; however, in electromagnetic compatibility, antenna theory, numerical calculation of electric and magnetic fields, electric energy systems, plasma physics, biocybernetics, etc. there is no escape. Despite their importance, many students are not as familiar with these equations as would be desirable; for many an electrical engineer they remain a sealed book all their lives. Therefore, this booklet attempts to introduce the reader gently on a *step-by-step basis* into field theory jargon and to present the essence of Maxwell's equations in a concise yet illustrative manner. In order to avoid that the reader "cannot see the forest for the trees," this unconventional introduction frequently quits where many established books on field theory elaborate; this is part of this booklet's appeal. Many years of teaching, research and development in several of the fields mentioned above have prompted the author to choose this presentation, which by no means is intended to replace proven textbooks, rather to lure students into reading those. It is assumed, however, that the reader has some familiarity with electricity and magnetism.

Even very elementary terms, e.g. *flux, induction, displacement* etc. make beginners feel lastingly uncomfortable. Therefore, this text starts with a demonstration of obvious formal analogies of electric, magnetic and conduction field quantities. This is followed by a detailed interpretation of Maxwell's equations in integral form.

The differential form of Maxwell's equations calls for a comprehensive explanation of the terms *curl* and *div*, which generally are conceptually demanding. The consistent use of the descriptive paired concepts *vortex strength* and *vortex density* for circulation and rotation (curl), as well as of the paired concepts *source strength* and *source density* for the net flux through a closed surface and for the divergence, is regarded an innovative contribution to a thorough understanding of the relationship between *global* and *local* field quantities. Although, at first glance, those paired concepts may look strange to the reader they resemble the fundamental idea of a transparently structured electromagnetic field theory.

The explanation of the terms *gradient, potential,* and *potential function* is followed by the *potential equations* for electric fields with and without space charges. Likewise, the deduction of scalar and vector magnetic potentials is followed by the scalar and vector potential equations for magnetic fields. The discussion of scalar and vector potential concepts serves simultaneously as the first introduction of the inverse integral operators $\{curl\}^{-1}$, $\{div\}^{-1}$ and $\{grad\}^{-1}$ which prove as helpful tools in finding the general solution of differential equations of several independent variables. When electric and magnetic fields are classified according to their time dependency, considerable emphasis is placed on an intimate understanding of the terms *static, quasi-static* and *nonstationary* fields, and on the deduction of the wave equation.

The easily understood transmission-line equations provide the basis for a systematic, distinct classification of terms such as *telegraphist's* equation, *wave* equation, *diffusion* equation, Laplace equation, Helmholtz equation, and the renowned Schroedinger equation. This is followed by Lorentz's *invariance* of Maxwell's equations which demonstrates that the difference between electric and magnetic fields is not as marked as daily engineering problems would lead one to believe.

Finally, the last chapter aquaints the reader with four different concepts of numerical field calculation – *finite-element* method, *finite-difference* method, *charge simulation* method and the Monte Carlo method.

The author is grateful to former and present students A. and I. Brauch, R. Büche, P. Deister, Th. Dunz, Dr. F. Imo, H. Kunz, R. Maier, B. Schaub, and Dr. H.H. Zimmer for their comments and proof reading. May thanks also extend to many anonymous reviewers who have inspired me to make changes in the second German edition.

Special thanks go to Mrs. Therese Färber for her contributions during the translation, to Mrs. Madeleine Michels and Mrs. Gerdi Ottmar, appreciated typists of the camera-ready manuscript, to our artist Mrs. Edith Müller, and ultimately, to Dr. Julia Abrahams, Albrecht von Hagen, and Dr. Ludwig, Springer publishers, for their steady support and encouragement.

Anyone finding errors in the text or figures is kindly invited to write to the author for the benefit of the readers of a subsequent printing.

Karlsruhe, June, 1988

Prof. Dr. Adolf J. Schwab

High-Voltage Research Laboratory
University of Karlsruhe
7500 Karlsruhe
FR-Germany

Contents

1 Elementary Concepts of Electric and Magnetic Fields. **1**

1.1 Flux and Flux Density of Vector Fields 3

1.2 Equations of Matter - Constitutive Relations 10

2 Types of Vector Fields **15**

2.1 Electric Source Fields 15

2.2 Electric and Magnetic Vortex Fields 16

2.3 General Vector Fields 17

3 Field Theory Equations **19**

3.1 Integral Form of Maxwell´s Equations 19

3.1.1 Faraday´s Induction Law in Integral Form
Vortex Strength of Electric Vortex Fields 20

3.1.2 Ampere´s Circuital Law in Integral Form
Vortex Strength of Magnetic Vortex Fields 23

3.1.3 Gauss´s Law of the Electric Field
Source Strength of Electric Fields 30

3.1.4 Gauss´s Law of the Magnetic Field
Source Strength of Magnetic Fields 31

3.2 Law of Continuity in Integral Form
Source Strength of Current Density Fields 32

3.3 Differential Form of Maxwell´s Equations 37

3.3.1 Faraday´s Induction Law in Differential Form *Vortex Density* of Electric Vortex Fields 38

3.3.2 Ampere´s Circuital Law in Differential Form *Vortex Density* of Magnetic Vortex Fields 42

3.3.3 Divergence of Electric Fields *Source Density* of Electric Fields 44

3.3.4 Divergence of Magnetic Fields *Source Density* of Magnetic Fields 46

3.4 Law of Continuity in Differential Form *Source Density* of Current Density Fields 47

3.5 Maxwell´s Equations in Complex Notation 52

3.6 Integral Theorems of Stokes and Gauss 53

3.7 Network Model of Induction 55

4 Gradient, Potential, Potential Function 60

4.1 Gradient of a Scalar Field 62

4.2 Potential and Potential Function of Static Electric Fields 65

4.3 Development of the Potential Function from a Given Charge Distribution 69

4.3.1 Potential Function of a Line Charge 72

4.3.2 Potential Function of a General Charge Distribution 75

4.4 Potential Equations 77

4.4.1 Potential Equations for Fields without Space Charges 77

4.4.2 Potential Equations for Fields with Space Charges 80

4.5 Electric Vector Potential 84

4.6 Vector Potential of the Conduction Field 87

5 Potential and Potential Function of Magnetostatic Fields ..90

5.1 Magnetic Scalar Potential 90

5.2 Potential Equation for Magnetic Scalar Potentials 92

5.3 Magnetic Vector Potential 94

5.4 Potential Equation for Magnetic Vector Potentials 100

6 Classification of Electric and Magnetic Fields 103

6.1 Stationary Fields 107

6.1.1 Electrostatic Fields 107

6.1.2 Magnetostatic Fields 109

6.1.3 Static Conduction Field (DC Current-Conduction Field) 111

6.2 Quasi-Stationary Fields (Steady-State) Fields 115

6.2.1 Quasi-Static Electric Fields 115

6.2.2 Quasi-Static Magnetic Fields 118

6.2.3 Quasi-Static Conduction Fields 120

6.2.4 Conduction Fields with Skin Effect 121

6.3 Nonstationary Fields, Electromagnetic Waves 125

6.3.1 Wave Equation 125

6.3.2 Retarded Potentials. 129

6.3.3 Hertz Potentials. 134

6.3.4 Energy *Density* in Electric and Magnetic Fields, Energy Flow *Density* in Electromagnetic Waves 137

7 Transmission-Line Equations **139**

8 Typical Differential Equations of Electrodynamics and Mathematical Physics **150**

8.1 Generalized Telegraphist´s Equation. 150

8.2 Telegraphist´s Equation with a, b>0; c=0 151

8.3 Telegraphist´s Equation with a>0; b=0; c=0 153

8.4 Telegraphist´s Equation with b>0; a=0; c=0 154

8.5 Helmholtz Equation 156

8.6 Schroedinger Equation 160

8.7 Lorentz´s Invariance of Maxwell´s Equations 163

9 Numerical Calculation of Potential Fields **169**

9.1 Finite-Element Method 169

9.2 Finite-Difference Method. 181

9.3 Charge Simulation Method 186

9.4 Monte Carlo Method. 190

9.5 General Remarks on Numerical Field Calculation 192

Appendix

A1 Units 194

A2 Scalar and Vector Integrals 197

A3 Vector Operations in Special Coordinate Systems 199

A4 Integral Operators $\{curl\}^{-1}$, $\{div\}^{-1}$, and $\{grad\}^{-1}$ 203

A5 Complex Notation of Harmonic Quantities 211

Literature. 213

Index. 216

1 Fundamental Terms of Electric and Magnetic Fields

Depending on the point of view, a field can be thought of as

a scalar or vector function of space,

a volume holding energy, or

a finite or infinite set whose elements are assigned unique values of a certain quantity, for example the point continuum of the Euclidian space, etc.

In any case, typical examples for *scalar fields* are the temperature distribution T(x,y,z) in a living-room or the potential function $\varphi(x,y,z)$ in the gap between charged electrodes; for *vector fields*, examples include the flow velocity $\mathbf{v}(x,y,z)$ in the wake of a surfboard or the magnetic field strength $\mathbf{H}(x,y,z)$ surrounding a current-carrying conductor.

Scalar and vector fields can have the following analytical appearance in a Cartesian coordinate system

Scalar field $\varphi(x,y,z) = 3x^2\,y + 7xz$

Vector field $\mathbf{E}(x,y,z) = 5z^2\,y\mathbf{a}_x + 2x^2\,\mathbf{a}_y + xy^2\,z\,\mathbf{a}_z\ .$ (1-1)

Upon insertion of coordinates x_v, y_v, z_v the first equation assigns each field point $P(x_v, y_v, z_v)$ a certain scalar value $\varphi_v = \varphi(x_v, y_v, z_v)$.

The second equation assigns to each field point a certain field vector $\mathbf{E}_v=\mathbf{E}(x_v,y_v,z_v)$, where $\mathbf{a}_x$, $\mathbf{a}_y$, and $\mathbf{a}_z$ represent unit vectors.

Alternatively, the Cartesian coordinates of a field point P_v can be interpreted as coordinates of a *position vector* $\mathbf{r}_v$ whose tip coincides with P_v. Then, field vectors can be represented as a function of this position vector $\mathbf{r}$, e.g. $\varphi(\mathbf{r})$ and $\mathbf{E}(\mathbf{r})$, Figure 1.1.

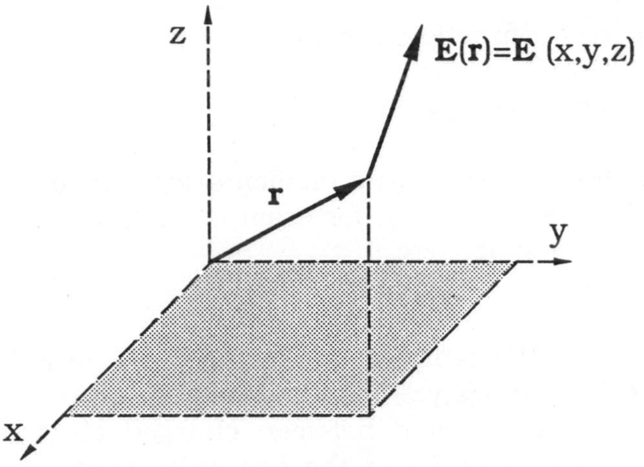

Figure 1.1: Representation of field vectors as a function of a position vector $\mathbf{r}$.

This representation is not restricted to a certain coordinate system; rather, $\mathbf{r}$ can represent any triple of numbers of an arbitrary three-dimensional coordinate system. Therefore, we will use this notation whenever we deal with the general physical nature of field quantities or field equations.

As a convenient starting point we will illustrate some elementary field theory concepts.

1.1 Flux Density of Vector Fields

An *electric field*, for instance, exists between the electrodes of a parallel-plate capacitor with area S, spacing d, and charges +Q and -Q.

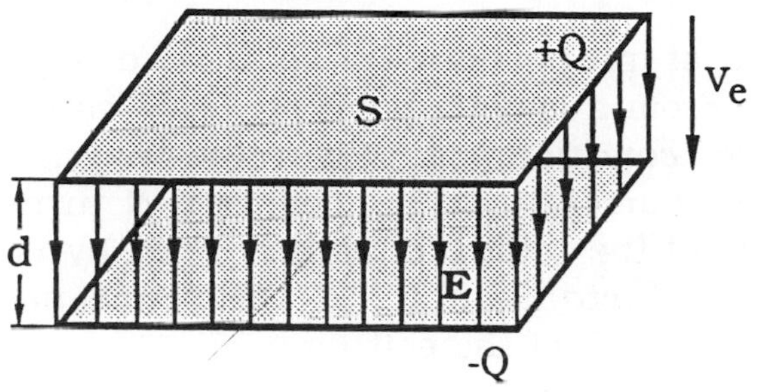

Figure 1.2: Electric field in an ideal parallel-plate capacitor.

Between both plates exists an *electric voltage* V_e which can be measured with a high-impedance voltmeter. The voltage V_e produces a proportional *electric flux* ψ between the electrodes (in terms of field theory the charges actually do this),

$$\boxed{\psi = C \cdot V_e} \qquad (1\text{-}2)$$

The constant of proportionality is designated as the *capacitance* C. From a generic point of view C can be interpreted as a *dielectric conductance.*

Inside the capacitor there exists an *electric field strength* $E=V_e/d$.

A *magnetic field* exists when the plates represent the pole pieces of a permanent magnet or a square sector of the magnetic gap between stator and rotor of an electrical machine.

Between both plates exists a *magnetic voltage* V_m producing a *magnetic flux* ϕ,

$$\boxed{\phi = P \cdot V_m} \qquad (1\text{-}3)$$

The constant of proportionality is here designated as the *permeance* P. From a generic point of view P can be interpreted as a *magnetic conductance.* When the magnetic voltage is generated by a current-carrying coil with N turns (see 3.1.2), the permeance of the total flux path is given by $P=L/N^2$, where L is the coil's *inductance*; for N=1 the permeance is identical with the inductance L of the coil.

In the magnetic gap between the sectors S exists a *magnetic field strength* $H=V_m/d$.

Electric and magnetic voltages possess only formal analogy; regarding their physical nature they are very different. For instance, the magnetic voltage has the dimensions ampere or, for N>1, ampere-turns (see 3.1.2).

If the volume between both plates is filled with conductive matter and if both plates are connected to a constant voltage source, a *conduction flux,* i.e. the familiar *electric current* I, is observed.

$$\boxed{I = G \cdot V_e} \qquad (1\text{-}4)$$

The field found inside conductors is called the *conduction field* or *J-field.* Its proportionality constant is designated as the *conductance G.* From a generic point of view G could be, more specifically, designated as the *ohmic conductance.*

Customarily, *flux densities* are used instead of fluxes because the flux quantity makes only an integral statement depending on both the area of the surface considered *and* the strength of the vector field penetrating that area. For instance, the flux of the uniform field of a parallel-plate capacitor ignoring fringe effects is given by

$$\psi = C\,V_e = \varepsilon \frac{S}{d} V_e = \varepsilon\, S \frac{V_e}{d} = \varepsilon\, S\, E \quad . \tag{1-5}$$

Hence, speaking of flux necessarily implies the existence of a certain surface **S** with area S penetrated by that flux (exception: flux of a point charge through a closed contour, see 3.1.3 and 3.1.4). Regarding electric fields we have, in general, $\psi=\psi(\varepsilon\mathbf{E},\mathbf{S})$ or $\psi=\psi(\mathbf{D},\mathbf{S})$, where the vector $\mathbf{S}=S\mathbf{n}_S$ represents a surface in space. The magnitude S specifies its area, the set of unit normal vectors $\mathbf{n}_S$ its orientation. If the area considered is not *eo ipso* perpendicular to the direction of flow, only its projection on a plane perpendicular to that direction is taken into account for the calculation of a flux. This is due to the nature of a scalar product as is discussed below.

The mere value of a particular flux does not indicate whether it is due to a large area or a strong vector field (as is the case with a given amount of work, which does not let us know what the relative contributions of force and distance may be).

In order to allow a statement about the vector field's contribution, the flux is related to the area considered. For instance, for an electric field this yields the *flux density* **D**. It would be convenient to simply divide the flux by the vector area. For example, the electric flux density in the uniform field of a parallel-plate capacitor would be obtained as

$$\mathbf{D} = \frac{\psi}{\mathbf{S}} \quad , \tag{1-6}$$

the space-dependent local flux density in a nonuniform electric field as

$$\mathbf{D} = \frac{d\psi}{d\mathbf{S}} \quad . \tag{1-7}$$

This definition, however, is multi-valued because, in general, an arbitrary scalar product

$$\psi = \mathbf{D} \cdot \mathbf{S} = DS \cos\alpha \tag{1-8}$$

cannot be solved uniquely for one of its two vectors. Only one equation exists for several unknowns (n coordinates of the unknown vector).

To say it in a different way, a scalar product can only be solved for the projections of one vector onto the other, i.e. $|\mathbf{D}| \cos\alpha$ or $|\mathbf{S}| \cos\alpha$, respectively. This is the reason why division by a vector is not defined in vector algebra.

An important exception where the unique inversion is obtainable is the case when the vectors **D** and **S**, or **D** and d**S**, are *eo ipso* presumed to be collinear. In this case the direction of the unknown vector is given, and only its magnitude needs to be determined. Due to the definition of a scalar product, (1-8), we obtain for $\alpha = 0$

$$D = \frac{\psi}{S} \quad \text{and} \quad D = \frac{d\psi}{dS} \quad . \tag{1-9}$$

These equations look familiar and can be found in many text-books. However, if one does not verbally imply collinearity and specify a particular direction of either **D** or **S**, these equations are as multi-valued as the inversion $\mathbf{D} = \psi/\mathbf{S}$ of a general scalar product. Thus, regarding uniqueness, Eqs. (1-9) do not deserve more respect than Eqs. (1-6) and (1-7). Moreover, what if **S** and d**S** are not *eo ipso* perpendicular, Figure 1.3 ?

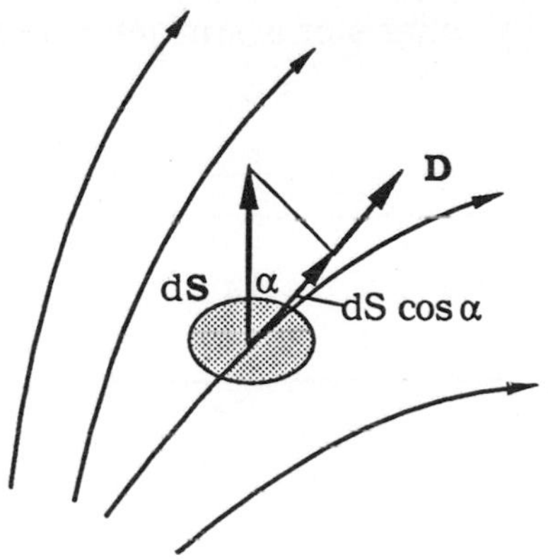

Figure 1.3 General case of non-collinear vectors $\mathbf{dS} = dS\,\mathbf{n}_S$ and $\mathbf{D} = D\,\mathbf{n}_D$

In this general case we must take the surfaces projections, i.e. preserve cosα,

$$D = \frac{\psi}{S\cos\alpha} \quad \text{and} \quad D = \frac{d\psi}{dS\cos\alpha} \quad . \qquad (1\text{-}10)$$

Yet, equations (1-10) do not define the vector **D** but only a scalar D. To overcome this problem, we multiply (1-10) by the unit vector $\mathbf{n}_D$ and finally obtain the desired definitions for the flux density *vector* **D** of uniform and nonuniform fields,

$$\boxed{\mathbf{D} = \frac{\psi}{S\cos\alpha}\mathbf{n}_D} \quad \text{and} \quad \boxed{\mathbf{D} = \frac{d\psi}{dS\cos\alpha}\mathbf{n}_D} \quad . \qquad (1\text{-}11)$$

Equations (1-11) look more complex than (1-6) and (1-7). However, in general, most engineering problems imply α=0.

Then equations (1-11) reduce to simple ratios of the type $D=\psi/A$.

From a given flux density and area the associated flux can be calculated any time by means of a surface integral,

$$\boxed{\psi = \int_S \mathbf{D} \cdot d\mathbf{S}} \quad . \tag{1-12}$$

In this instant, the reader should not try to rack his or her brain about how to analytically evaluate a surface integral. For the moment it is entirely sufficient to know that the result of multiplying a flux density by a surface element has the dimension of flux and that the integral merely sums up infinitesimal scalar products of local flux densities and their associated surface elements $d\mathbf{S}$ to give the total flux ψ of the integration region. Because of the definition of a scalar product - here $\mathbf{D} \cdot d\mathbf{S} = DdS \cos\alpha$ - only the projection of $d\mathbf{S}$ on a plane perpendicular to the respective flux density vector $\mathbf{D}$ is taken into account.

As shown above for the electric field, one can define in a magnetic field a *magnetic flux density*

$$\boxed{\mathbf{B} = \frac{\phi}{S \cos\alpha} \mathbf{n}_B} \quad \text{or} \quad \boxed{\mathbf{B} = \frac{d\phi}{dS \cos\alpha} \mathbf{n}_B} \tag{1-13}$$

and an associated magnetic flux

$$\phi = \int_S \mathbf{B} \cdot d\mathbf{S} \quad , \tag{1-14}$$

and in a conduction field an *electric current (conduction flux) density*

$$\mathbf{J} = \frac{I}{S \cos\alpha} \mathbf{n}_J \quad \text{or} \quad \mathbf{J} = \frac{dI}{dS \cos\alpha} \mathbf{n}_J \tag{1-15}$$

and an associated conduction flux

$$I = \int_S \mathbf{J} \cdot d\mathbf{S} \quad . \tag{1-16}$$

Generally, any surface integral over a vector quantity is called flux. This integral was first encountered in fluid mechanics, where actual matter flows. Regarding the electric flux ψ and the magnetic flux ϕ, nothing is effectively flowing. Nonetheless, we speak of flux because of the obvious formal analogy. Due to the different dimensions of their input quantities, ψ and ϕ possess dimensions which have nothing in common with the ordinary colloquial flux. Flux is a scalar quantity, therefore, the surface integral over a vector is called a scalar integral (see A2). However, note that a flux, while it is a scalar, always has associated with it a surface **S** to which it is functionally related by a flux integral (exception: a point charge´s flux through a closed surface, Gauss´s law; see 3.1.3).

Flux densities are used whenever a flux is not confined within specified bounds or when *local* statements about individual field points of a continuum are called for. Typical problems are conduction fields in electrolytes, eddy currents in conductors, saturation in magnetic circuits, stress in dielectrics, magnetic gap of electrical machines, far field of electromagnetic waves, etc.

Because flux can only be assigned to an area, rather than to a discrete point – a function of position, e.g. $\psi(\mathbf{r})$ or $\psi(x,y,z)$ does not exist (exception: a point charge's total flux) – flux must be graphically represented by *flux tubes* (tubes of constant flux). To avoid this difficulty, graphic presentation of a flux by its flux density is generally preferred, allowing the portrayal of flux characteristics by field lines. The flux density vector $\mathbf{D}(\mathbf{r}_V)$ at a field point $P(\mathbf{r}_V)$ is collinear with the tangent to the field line through $P(\mathbf{r}_V)$. With sufficiently fine discretization, field lines represent the central axes of the flux tubes.

The terms *displacement* or *displacement density* for the electric flux density **D**, and *induction* for the magnetic flux density **B** should be utilized only when the reader has acquired a thorough understanding of the *flux* and *flux-density* concept. Actually, those terms are entirely dispensable and merely historic. (So far field theory has not greatly felt the absence of a like term for the conduction-current flux density **J**). Therefore, it is recommended that the reader use the generic *flux-density concept,* i.e. electric, magnetic, and conduction flux-densities.

1.2 Equations of Matter – Constitutive Relations

Dividing equations (1-2), (1-3), and (1-4) of section 1.1 by the area S yields *flux densities* on the left sides and *field strengths* on the right sides, related through their respective relative conductivities ε, μ, or σ. These equations of matter are referred to as *constitutive relations.*

$$\boxed{\mathbf{D} = \varepsilon\mathbf{E} \qquad \mathbf{B} = \mu\mathbf{H} \qquad \mathbf{J} = \sigma\mathbf{E}} \quad . \qquad (1\text{-}17)$$

Here and in the following the concept of conductivity is frequently used in a generic sense, independent of a vector field´s nature (e.g. *thermal* conductivity).

In general, given a certain field strength the flux density will be higher the greater the considered medium´s conductivity is. The converse is also true, that is given a certain flux, a high conductivity will be associated with a low field strength, for example field strengths in the iron parts of magnetic circuits or in electrical conductors.

Flux densities and field strengths are proportional only in homogeneous, isotropic, linear matter. Otherwise the factor of proportionality can be a function of space or the field vectors´ direction and magnitude. *Possible exceptions from proportionality are generally so obvious that the necessary presumptions cited above are not repeatedly claimed in the following text.*

At boundaries between two media 1 and 2, the normal components of the flux densities on both sides are continuous (if there are no sources of flux at that interface),

$$\boxed{D_{n1} = D_{n2} \qquad B_{n1} = B_{n2} \qquad J_{n1} = J_{n2}} \quad . \qquad (1\text{-}18)$$

In many technical applications an electric or magnetic voltage drives an electric, magnetic or conduction flux (current) through a flux path. If the path consists of serveral series-connected media of different conductivities, the normal components of the flux densities possess equal magnitudes on both sides of a boundary. Moreover, when a uniform cross-section is maintained all normal components will be constant along the entire flux path.

The field strength in a particular medium depends on its conductivity and, for a given flux density, is determined by the equations of matter (constitutive relations), Figure 1.4.

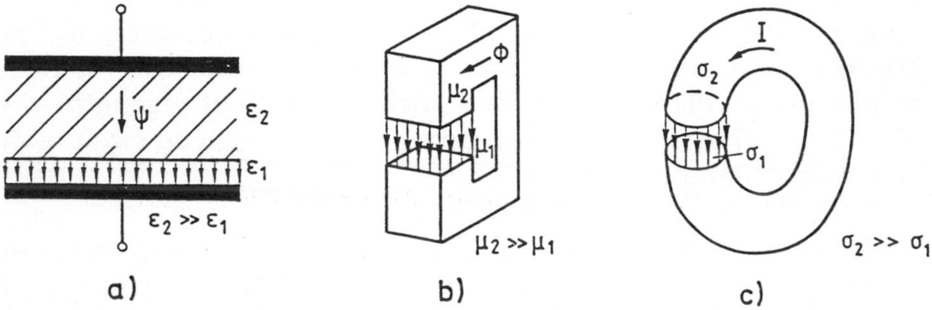

Figure1.4: Distribution of field strengths according to various conductivities along flux paths (flux driving sources are not drawn).
a) capacitor with gas-filled gap
b) magnetic circuit with air gap
c) open transformer winding (see 3.1.1)

This explains why in multi-layer dielectrics the material with the least permittivity suffers the highest electric stress (e.g. glow discharges in gas-filled cavities inside solid dielectrics), why the maximum magnetic field strength of a magnetic circuit occurs in its magnetic gap, and why an induced voltage can be measured between the ends of an open winding of a transformer (see 3.1.1).

In contrast to the continuous normal components of flux densities, the normal components of field strengths suffer a *step-change* at each boundary. For example, consider two dielectrics 1 and 2 in an electric field with flux densities $D_{n1} = \varepsilon_1 E_{n1}$ and $D_{n2} = \varepsilon_2 E_{n2}$. Since $D_{n1} = D_{n2}$, we have

$$\boxed{\frac{E_{n1}}{E_{n2}} = \frac{\varepsilon_2}{\varepsilon_1}} \quad . \tag{1-19}$$

In general, for electric, magnetic, and conduction fields we have

$$\boxed{E_{n1} = \frac{\varepsilon_2}{\varepsilon_1} E_{n2} \qquad H_{n1} = \frac{\mu_2}{\mu_1} H_{n2} \qquad E_{n1} = \frac{\sigma_2}{\sigma_1} E_{n2}} \quad . \tag{1-20}$$

The behaviour of field quantities on boundaries can be formally derived employing Gauss´s law for the respective fields (see 3.1.3, 3.1.4 and 3.2).

As a corollary, Figure 1.5 illustrates generically the boundary behaviour of *field-strength- and flux-density*-vectors. The field points on the boundary´s two sides are assumed to lie infinitesimally close together; duc to the limited graphical resolution they effectively coincide.

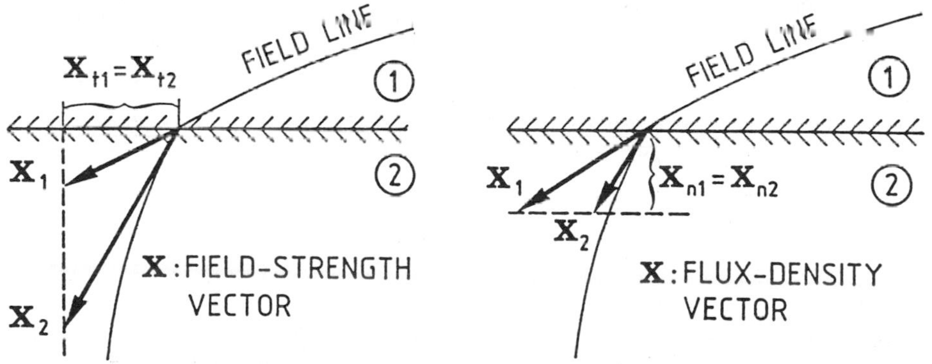

Figure 1.5: Boundary behaviour of general field strength and flux density vectors. The first possess continuous tangential components, the latter continuous normal components. The associated field lines suffer refraction.. (Regarding continuous tangential components, refer to the literature).

Finally, Table 1.1 provides a comparative survey of the terms and concepts discussed so far. The columns for nonuniform fields exhibit integral equations taking into account the space dependency of field quantities in nonuniform fields. These equations will be explained in more detail at a later stage. They were included here only for the sake of completeness. For the moment, the reader needs only the equations for uniform fields.

PHYSICAL QUANTITY	E- FIELD		H - FIELD		J - FIELD (CONDUCTION FIELD)	
	uniform	nonuniform	uniform	nonuniform	uniform	nonuniform
VOLTAGE V	$V_e = Ed$ [V]	$V_e = \int \mathbf{E}\cdot d\mathbf{r}$ [V]	$V_m = Hd$ [A]	$V_m = \int \mathbf{H}\cdot d\mathbf{r}$ [A]	$V_e = Ed$ [V]	$V_e = \int \mathbf{E}\cdot d\mathbf{r}$ [V]
FIELD STRENGTH	$E = \frac{V_e}{d}$	$\mathbf{E} = \frac{dV_e}{d\mathbf{r}}$	$H = \frac{V_m}{d}$	$\mathbf{H} = \frac{dV_m}{d\mathbf{r}}$	$E = \frac{V_e}{d}$	$\mathbf{E} = \frac{dV_e}{d\mathbf{r}}$
FLUX	$\psi = CV_e$		$\phi = \Lambda V_m$, $\Lambda = \frac{L}{N^2}$ $(N=1 \rightarrow \Lambda = L)$		$I = GV_e$ (CONDUCTION CURRENT)	
FLUX-DENSITY	$\mathbf{D} = \frac{\psi}{S\cos\alpha}\mathbf{n}_D$	$\mathbf{D} = \frac{d\psi}{dS\cos\alpha}\mathbf{n}_D$	$\mathbf{B} = \frac{\phi}{S\cos\alpha}\mathbf{n}_B$	$\mathbf{B} = \frac{d\phi}{dS\cos\alpha}\mathbf{n}_B$	$\mathbf{J} = \frac{I}{S\cos\alpha}\mathbf{n}_J$	$\mathbf{J} = \frac{dI}{dS\cos\alpha}\mathbf{n}_J$
	D: DISPLACEMENT		**B**: INDUCTION		**J**: CURRENT DENSITY	
	$\psi = \mathbf{D}\cdot\mathbf{S}$	$\psi = \int \mathbf{D}\cdot d\mathbf{S}$	$\phi = \mathbf{B}\cdot\mathbf{S}$	$\phi = \int \mathbf{B}\cdot d\mathbf{S}$	$I = \mathbf{J}\cdot\mathbf{S}$	$I = \int \mathbf{J}\cdot d\mathbf{S}$
		$\psi = \oint \mathbf{D}\cdot d\mathbf{S} = Q$		$\phi = \oint \mathbf{B}\cdot d\mathbf{S} = 0$		$I = \oint \mathbf{J}\cdot d\mathbf{S} = 0$
FIELD CONDUCTIVITY	$\mathbf{D} = \varepsilon\mathbf{E}$ ε: DIELECTRIC CONDUCTIVITY (PERMITTIVITY)		$\mathbf{B} = \mu\mathbf{H}$ μ: MAGNETIC CONDUCTIVITY (PERMEABILITY)		$\mathbf{J} = \sigma\mathbf{E}$ σ: (OHMIC) CONDUCTIVITY	

Table 1.1: Comparative survey of field quantities of electric and magnetic fields and of the conduction field. Later in the text, the current density **J** of the conduction field has the subscript "c" to distinguish it from the total current density (see 3.1.2).

2 Types of Vector Fields

Vector fields can be classified as

- *source fields* (synonymously called lamellar, irrotational, or conservative fields) and

- *vortex fields* (synonymously called solenoidal, rotational, or nonconservative fields)

Electric fields **E** (x,y,z) can be source or vortex fields, or combinations of both, while magnetic fields **B** (x,y,z) are always vortex fields (see 3.1.4). In this text the concept of *sources* is consistently reserved for positive or negative electric charges only. Vortex fields are said to originate from *vortices*.

2.1 Electric Source-Fields

Electric source-fields exist in the environment of electric charges at rest (electrostatic charges, fixed or localized charges). The existence of these fields can be demonstrated via their forces on objects brought into the field.

The field lines possess starting and terminating points, they originate and terminate on *sources*, these are positive and negative charges (sources with negative sign are frequently called *sinks)*. One can distinguish between pure space-charge problems, these are fields without boundaries (Newton potentials), and so-called boundary-value problems, these are fields between physical electrodes (boundaries), Figure 2.1 a, b.

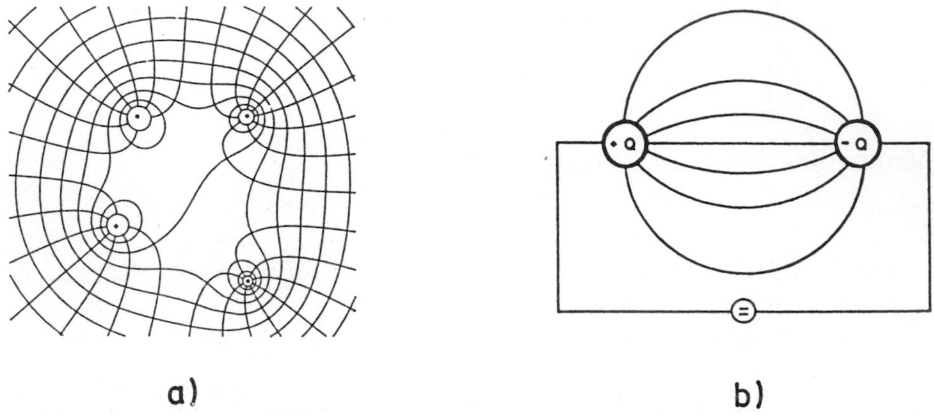

Figure 2.1: Electric source-fields of fixed electric charges
a) pure space-charge field without boundaries
b) field between electrodes (boundary-value problem).

One example of a real physical system where both types of problems exist is the situation of high electric field-strengths producing free charges in an electrode gap.

2.2 Electric and Magnetic Vortex Fields

Electric vortex-fields exist in the environment of a time-varying magnetic flux, for example inside and outside a transformer leg, Figure 2.2a. Magnetic vortex-fields exist in the environment of a time varying or constant current-flux, for example inside and outside conductors carrying ac or dc, Figure 2.2b.

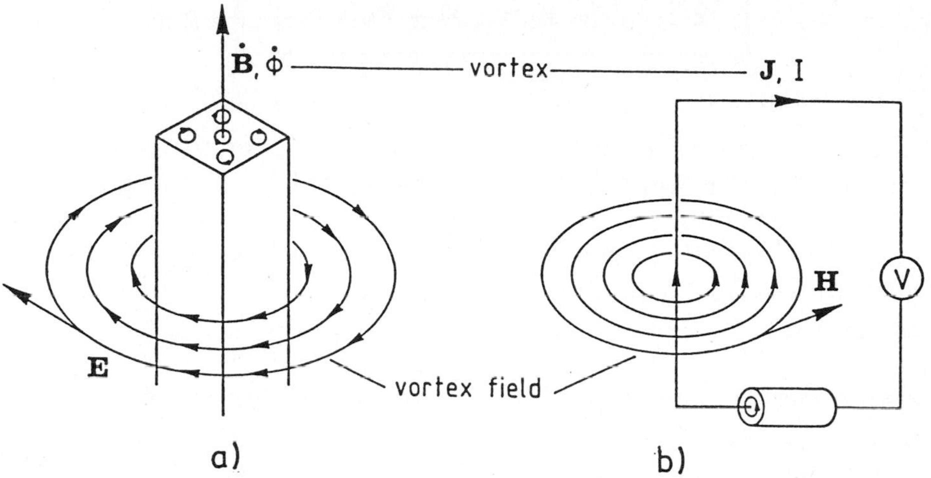

Figure 2.2: Examples of electric and magnetic vortex-fields; a) electric vortex-field, b) magnetic vortex-field, **E** and $\dot{\mathbf{B}}$ obey the left-hand rule, **H** and **J** the right-hand rule.

Field lines of vortex fields lack starting or terminating points; they are solenoidal. *Linear or tubular regions around which vortex-field lines contract are called vortices of the respective vortex field.* Hence, vortices of electric vortex-fields are $\dot{\phi}$ or $\dot{\mathbf{B}}$-lines, vortices of magnetic vortex-fields are I, **J**- or $\dot{\mathbf{D}}$- lines (considering the remarks in Chapter 1 regarding the different quality of flux tubes and lines). The vortices are also solenoidal. For instance, the flux lines and their time derivatives in a transformer core, or the current filaments in a circuit form solenoidal lines (no current flows in an open circuit).

2.3 General Vector Fields

In general, a vector field, e.g. **E**(x,y,z), can be composed of both a source field and a vortex field (Fundamental theorem of vector analysis, for example Fig. 6.3 in Section 6.2.1)

$$\boxed{\mathbf{E}(x,y,z) = \mathbf{E}_S(x,y,z) + \mathbf{E}_V(x,y,z)} \quad . \qquad (2\text{-}1)$$

Hence, an arbitrary vector field is, with respect to its physical nature (i.e. the individual contributions of both components), uniquely specified only if its sources *and* vortices can be identified, in other words, if its *source density* and *vortex density* are given. These terms will be explained in detail in Chapter 3.

Mathematical modeling of fields can be accomplished with *far-action therory* (action at a distance), that is with remote conductor quantities Q, V, I or with *near-action theory* (field theory), that is with local field quantities **E**, **D**, **H**, **B**, (for example fields of antennas and electromagnetic waves). The links between both types of quantities are given by *Maxwell's equations.*

3 Field Theory Equations

Maxwell´s equations describe electromagnetic fields and their interaction with matter. Generally, these equations are encountered in the time domain, either in integral or differential form. For the solution of practical field problems, their complex notation is often preferred (frequency-domain notation). We will first discuss the time-domain integral form, which for the present is more illustrative and which lends itself more easily to physical reasoning and thought experiments.

3.1 Integral Form of Maxwell´s Equations

INTEGRAL FORM	
$\oint_C \mathbf{E} \cdot d\mathbf{r} = -\frac{d\phi}{dt} = \mathring{V}_e$	$\oint_C \mathbf{H} \cdot d\mathbf{r} = I = \mathring{V}_m$
$\oint_S \mathbf{D} \cdot d\mathbf{S} = Q$	$\oint_S \mathbf{B} \cdot d\mathbf{S} = 0$

Table 3.1: Integral form of Maxwell´s equations.

When this system of equations is encountered for the first time, it is not unusual for students to feel a sort of resignation and concern whether they will ever understand these equations in depth. This fear, however, is unjustified, because, in general, only the physical essence of these equations need to be understood. Mathematical evaluation of the vector integrals is rarely asked for since real field problems are usually attacked using Maxwell´s equations in differential form. Frequently, solutions of Maxwell´s equations - these are the electric and magnetic field-strengths **E**(**r**) and **H**(**r**) - are obtained via mathematical surrogate functions, referred to as *potentials* (see Chapter 4).

We will now discuss the individual equations in more detail and begin with the first row.

3.1.1 Faraday´s Induction Law in Integral Form *Vortex Strength* of Electric Vortex-Fields

A time-varying magnetic flux, e.g. in a transformer leg, generates (induces) an electric voltage V_e in a conductor loop surrounding it. This voltage can be measured upon opening the loop, Figure 3.1 a).

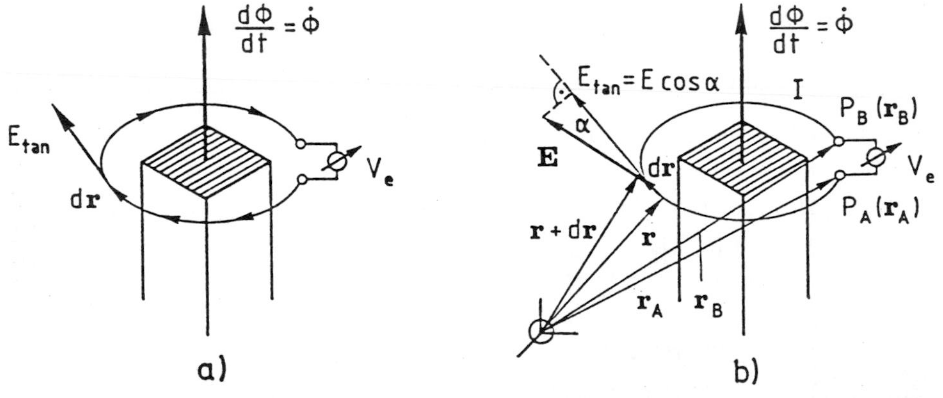

Figure 3.1: Transformer leg with time-varying flux and open circuited conductor loop. To avoid portraying flux tubes, the flux derivative $\dot{\phi}$ is represented by the derivative of its flux density, $\dot{\mathbf{B}}$. The quantities **E** and $\dot{\phi}$ obey the left-hand rule, **E** and -$\dot{\phi}$ or -$\dot{\mathbf{B}}$ the right-hand rule.

The *induction effect* is observed because a time-varying magnetic flux ϕ is surrounded by an electric vortex-field with solenoidal field lines in much the same way as a current flux I is surrounded by a magnetic vortex-field. Recall that vortices of electric vortex-fields are $\dot{\phi}$ or $\dot{\mathbf{B}}$-lines, vortices of magnetic vortex-fields are I or $\mathbf{J}$-lines (see 2.2).

Multiplying the tangential component of the field strength of the electric vortex-field by the associated path element dr yields the infinitesimal voltage induced across dr,

$$dV_e = E_{tan}\ dr \quad . \tag{3-1}$$

Formally, dV_e is obtained as the result of the scalar product of the field-strength vector $\mathbf{E}$ and the path element $\mathbf{dr}$, Fig. 3.1b,

$$dV_e = \mathbf{E} \cdot \mathbf{dr} = E \cos\alpha\ dr = E_{tan}\ dr \quad . \tag{3-2}$$

Summing up all infinitesimal voltages dV_e around the conductor loop from $\mathbf{r}_A$ to $\mathbf{r}_B$ yields the voltage induced between the terminals $P_A(\mathbf{r}_A)$ and $P_B(\mathbf{r}_B)$.

$$V_e = \int_{\mathbf{r}_A}^{\mathbf{r}_B} \mathbf{E} \cdot \mathbf{dr} \quad . \tag{3-3}$$

If the terminals $P_A(\mathbf{r}_A)$ and $P_B(\mathbf{r}_B)$ come very close and eventually coincide (forming a closed loop), then, the voltage V_e induced in the closed contour becomes the *circulation voltage* $\mathring{V}_e$, referred to as the *electromotive force "emf "*

$$\boxed{\mathring{V}_e = \oint_C \mathbf{E} \cdot \mathbf{dr} = -\frac{d\phi}{dt} = -\dot{\phi}} \quad , \tag{3-4}$$

where the circle superscript of the voltage symbol and the circle in the integral symbol indicate the closed integration path. The induced circulation voltage is proportional to the negative time rate-of-change of the magnetic flux and, therefore, is a measure for the strength of the vortices of the electric field, $\dot{\phi}$ or $\dot{\mathbf{B}}$ in the magnetic circuit. Hence, it is called the *vortex strength of the electric vortex-field.* At first glance the paired concepts *vortex strength* and *vortex density* (3.3.1) may look strange to the reader, however, these concepts together with their counterparts *source strength* (3.1.3) and *source density* (3.3.3) are essentials of a transparently structured electromagnetic field theory.

If no time-varying magnetic flux penetrates the contour, the induced circulation voltage is zero. It is to be noted that the induction effect is not transferred magnetically but electrically. The electric field's vortices are confined in the transformer core; outside the core, there exists only the associated electric vortex-field.

The negative sign results from Lenz's law. The induced field strength is always directed such that the magnetic field of a current flowing in an imaginary conductor loop counteracts a change of the generating magnetic field of the magnetic circuit, that is it objects to a *variation* of state. **E** and $-\dot{\phi}$ are related by the right-hand rule (the direction of ϕ is arbitrary).

The great practical importance of the circulation voltage is due to the fact that it allows the calculation of the voltages in electrical machine windings from $\dot{\phi}$.

In its role as a circuit element, the induced voltage in an imaginary conductor loop corresponds to an ideal source; it is not short-circuited by the conductor loop. Depending on the loop's conductivity, arbitrarily high currents may flow (presuming that the initiating $\dot{\mathbf{B}}$ can be kept constant!).

Customarily, the flux ϕ is expressed by its flux density **B** via

$$\phi = \int \mathbf{B} \cdot d\mathbf{S}.$$

Then, Faraday's law takes the following form

$$\boxed{\overset{\circ}{V}_e = \oint_C \mathbf{E} \cdot d\mathbf{r} = -\frac{d}{dt} \int_S \mathbf{B} \cdot d\mathbf{S}} \qquad (3\text{-}5)$$

For conductors at rest, the sequence of the differential and integral operations is immaterial, thus one can write

$$\boxed{\overset{\circ}{V}_e = \oint_C \mathbf{E} \cdot d\mathbf{r} = -\int_S \frac{\partial \mathbf{B}}{\partial t} \cdot d\mathbf{S}} \qquad (3\text{-}6)$$

Furthermore, solenoidal electric field lines exist inside an iron core, where, because of the iron's conductivity, they cause *eddy currents*. In order to prevent or rather reduce them to negligible magnitudes (losses, dissipation), magnetic circuits are laminated, that is they are composed of many thin iron sheets insulated (electrically isolated) from each other.

3.1.2 Ampere's Circuital Law in Integral Form *Vortex Strength* of Magnetic Vortex Fields

A current flowing in a conductor loop drives a solenoidal magnetic flux *through* the area of this loop. Hence, current paths are surrounded by magnetic vortex-fields, Figure 3.2a. Just as current flux is guided by a conductor, so a magnetic flux can be guided by a magnetic circuit offering high magnetic conductivity (permeability) for magnetic flux, Figure 3.2 b.

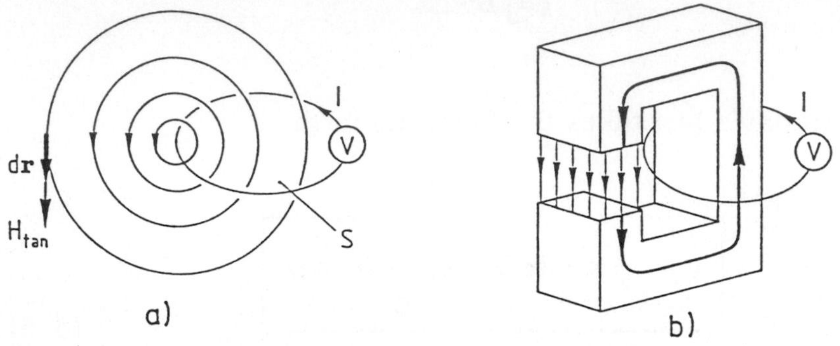

Figure 3.2: Magnetic vortex-field through a current loop, with and without magnetic circuit (schematic). I and **H** obey the right-hand rule.

In analogy to the electric field, the local strength of a magnetic field can be characterized by the magnetic field-strength **H** of the vortex field.

Multiplying the tangential component of the field strength of the magnetic field by a path element dr yields the infinitesimal magnetic voltage across dr (see 3.1.1)

$$dV_m = \mathbf{H} \cdot \mathbf{dr} = H \cos\alpha \; dr = H_{tan} \, dr \quad . \qquad (3\text{-}7)$$

Summing up all infinitesimal voltages dV_m around a closed contour C yields the *magnetic circulation voltage* $\mathring{V}_m$, referred to as *magnetomotive force "mmf"*,

$$\boxed{\mathring{V}_m = \oint_C \mathbf{H} \cdot \mathbf{dr} = I} \quad . \qquad (3\text{-}8)$$

For *one* turn the magnetic circulation-voltage $\mathring{V}_m$ is identical with the current I and thus is a measure for the strength of the vortices of the magnetic field (current-flux lines I). Hence, it is called *vortex strength of the magnetic vortex-field.* For N turns (*winding, coil*) the magnetic circulation-voltage equals the number of ampere-turns $N \cdot I = \theta$. The magnetic voltage $\mathring{V}_m$ has the dimensions ampere or ampere-turns (for $N>1$).

The great practical importance of the magnetic circulation voltage lies in the fact that it can be readily calculated from the total current-flux N·I and that it allows evaluation of the magnetic field-strength **H** upon division by the length of some flux path. Recall that field-strengths, in general, have the dimensions electric or magnetic voltage per distance. Specifically in the case of the magnetic field, the dimensions are expressed as ampere turns per distance.

The area S enclosed by the current loop determines the magnetic flux that will be driven through the contour by an ideal current source or the equivalent magnetic circulation-voltage $\mathring{V}_m$. In its role as an element of a magnetic circuit the magnetic circulation-voltage V_m corresponds to an ideal source that is not short-circuited by a high-conductance magnetic path.

Solenoidal magnetic field lines also surround individual current filaments in the interior of a conductor. For time-varying currents, the filaments' magnetic field is also varying, resulting in an electric vortex-field in the interior of the conductor. Due to the existing conductivity, this electric vortex field causes eddy currents, whose direction opposes the ordinary current flow in the center-axis and coincides with it at the circumference. Thus the current appears to be displaced from the center towards the periphery. This phenomenon is called *skin effect.* Skin effect is more pronounced the higher the frequency (for harmonic currents) or the higher the time rate-of change di/dt (non-sinusoidal wave-forms). It is to be noted that the skin effect is a linear phenomenon, that is it does not depend on current magnitude.

It must be emphasized that Ampere´s law considers the *total current,* i.e. the sum of *conduction current* I_c and *displacement current* I_d.

Basically, an electric current or its current density always consist of two components due to conduction and displacement,

$$\mathbf{J} = \underbrace{\mathbf{J}_c}_{\text{conduction current density}} + \underbrace{\mathbf{J}_d}_{\text{displacement current density}} = \sigma\mathbf{E} + \varepsilon\dot{\mathbf{E}} \quad . \qquad (3\text{-}9)$$

In the preceding sections **J** represented exclusively the conduction current density. In the following sections a conduction current density will be explicitly characterized by assigning **J** the subscript "c", while **J** without subscript is understood as the *total current density.*

Conduction currents result from moving electric charges, displacement currents from time-varying electric fields – $\varepsilon\dot{\mathbf{E}}$ or $\dot{\mathbf{D}}$. Frequently, the symbol **J** is reserved for the conduction component.

In conductors **D** is negligible compared with the conduction current density $\mathbf{J}_c$ up to frequencies in the optical region. In dielectrics the conduction current density is negligible compared with $\dot{\mathbf{D}}$,

$$\begin{array}{lllll} \text{conductor} & |\mathbf{J}_c| \gg |\mathbf{J}_d| & = & |\dot{\mathbf{D}}| \\ \text{dielectric} & |\mathbf{J}_c| \ll |\mathbf{J}_d| & = & |\dot{\mathbf{D}}| \\ \text{vacuum} & \mathbf{J}_c & = & 0 \quad . \end{array}$$

Figure 3.3 exhibits a section of a current loop interrupted by a dielectric gap.

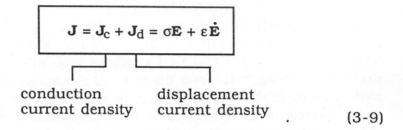

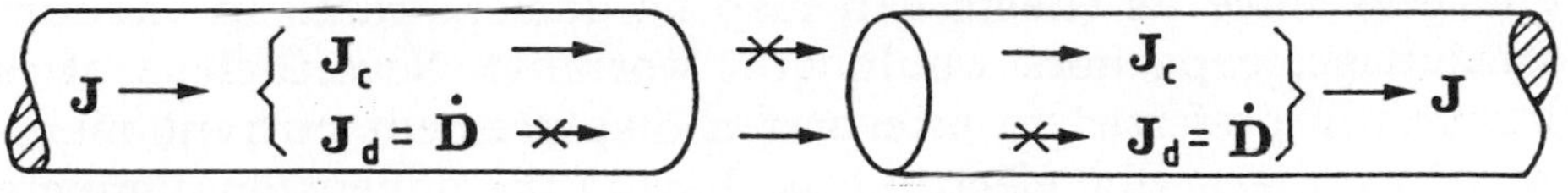

Figure 3.3: Illustration of the total electric current in a circuit with dielectric gap.

At the front ends of both conductors the normal component of the total current is continuous; conduction current readily turns into displacement current and vice versa. For sinusoidal quantities the conduction current in the conductor and its continuation as displacement current in the dielectric are in phase. Within the same medium they possess a phase-shift of 90°.

For complex quantities (see A5) a frequency limit ω_L can be defined, above or below which the displacement current density can be neglected over the conduction current density

$$\boxed{\underline{\mathbf{J}} = \sigma\underline{\mathbf{E}} + j\omega\varepsilon\underline{\mathbf{E}} = (\sigma + j\omega\varepsilon)\ \underline{\mathbf{E}}} \quad . \tag{3-10}$$

Then, independently of field-strength magnitude, we find

$\sigma \gg \omega\varepsilon$	conduction current dominant
$\sigma \ll \omega\varepsilon$	displacement current dominant.

Whereas typical good conductors (e.g. copper or aluminum) and typical good dielectrics (e.g. PVC or PE) maintain their characteristic properties practically independent of frequency, midway between these two, many materials such as soil, seawater, or bio-organisms can not be generally classified. The frequency of the imposed field will determine whether a medium is considered a conductor or a lossy dielectric.

Regarding discussions as to whether displacement currents are real, one can say that displacement current positively exists, though not necessarily in its original sense (displacement of charges) since its effects can also be demonstrated in vacuum (otherwise, capacitors would quit working). Nevertheless, it is frequently preferred to interpret a displacement current merely as a time-varying electric flux, having the dimensions ampere in much the same way as a conduction current.

Customarily, the total current flux I is expressed by the flux density of the total electric current

$$I = \int_S \mathbf{J} \cdot d\mathbf{S} = \int_S (\mathbf{J}_c + \dot{\mathbf{D}}) \cdot d\mathbf{S} \quad , \tag{3-11}$$

then Ampere´s law becomes

$$\boxed{\overset{\circ}{V}_m = \oint_C \mathbf{H} \cdot d\mathbf{r} = \int_S (\mathbf{J}_c + \dot{\mathbf{D}}) \cdot d\mathbf{S}} \quad . \tag{3-12}$$

To summarize, the left sides of the first row of Maxwell´s equations in Table 3.1 possess great formal similarity and represent a particular type of integral which is encountered in many other disciplines as well. This type of integral is generically designated as *vortex strength*. Other possible terms are circulation voltage or electro- or magnetomotive force.

For any vector field **X**:

$$\boxed{\oint_C \mathbf{X} \cdot d\mathbf{r} = \begin{cases} \textit{vortex strength} \\ \text{motive force} \\ \text{circulation} \end{cases}} \quad , \tag{3-13}$$

the terms on the right are synonymous.

$\oint \mathbf{X} \cdot d\mathbf{r} = 0$: The contour C is vortex-free, or the sum of all vortices through the contour equals zero.

$\oint \mathbf{X} \cdot d\mathbf{r} \neq 0$: The contour C comprises vortices, the net vortex strength is nonzero.

As a corollary,

$$\oint_C \mathbf{X} \cdot d\mathbf{r} = 0 \qquad (3\text{-}14)$$

in a limited region does not indicate that we deal with a source field, only that the region considered does not contain a net vortex strength, Figure 3.4.

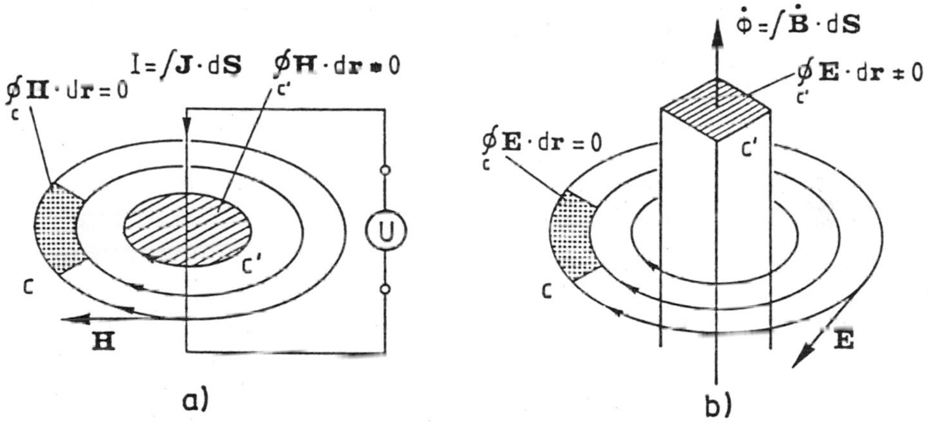

Figure 3.4: Regions of vortex fields with and without vortices a) magnetic field, b) electric field.

If a problem implies *eo ipso* a pure source field $\mathbf{X}_S(\mathbf{r})$, we always have

$$\oint_S \mathbf{X}_S(\mathbf{r}) \cdot d\mathbf{r} = 0. \qquad (3\text{-}15)$$

3.1.3 Gauss´s Law of the Electric Field *Source Strength* of Electric Fields

As has been shown in Chapter 2, the electric flux ψ penetrating a surface can be calculated from its flux density **D** and the area S of the surface **S**, Figure 3.5 a.

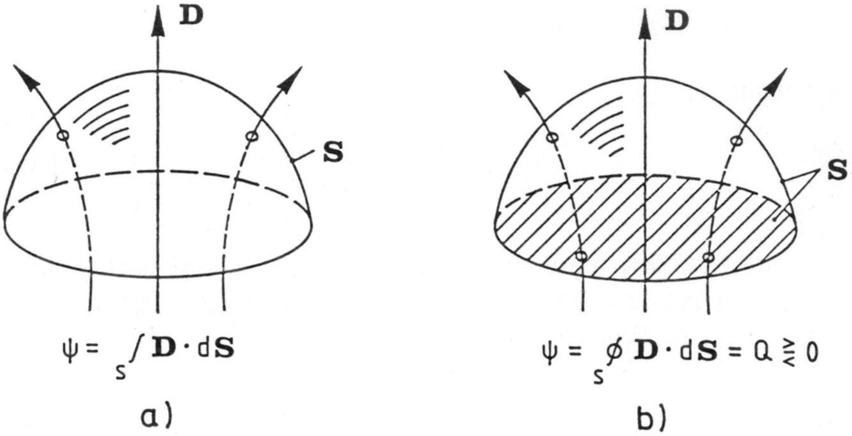

Figure 3.5: Gauss´s law of the electric field
a) open surface, b) closed surface.

Considering an arbitrary closed surface, for example the curved surface in Figure 3.5b together with its dashed bottom, the integral equals zero if the closed surface does not contain a net charge. If the integration yields a non-zero flux, this flux is identical with the contained charge or a surplus of charges of one polarity. In the first case where the integral is zero, we are either talking about the source-free region of a source field or about a vortex field. The flux entering the closed surface equals the flux leaving the surface. The simple example of water flowing through a garden hose illustrates this statement very clearly.

If the integral yields a non-zero flux, the integration region considered contains a source or a source field (irrespective of any additionally existing electric vortex-field). The net flux ψ=Q penetrating a closed surface is called the *source strength of the electric field* of the region considered.

3.1.4 Gauss's Law of the Magnetic Field *Source Strength* of Magnetic Fields

As has been shown in Chapter 2, the magnetic flux ϕ penetrating the surface of a current loop C can be calculated from the flux density **B** and the area S of the surface **S**, Figure 3.6a.

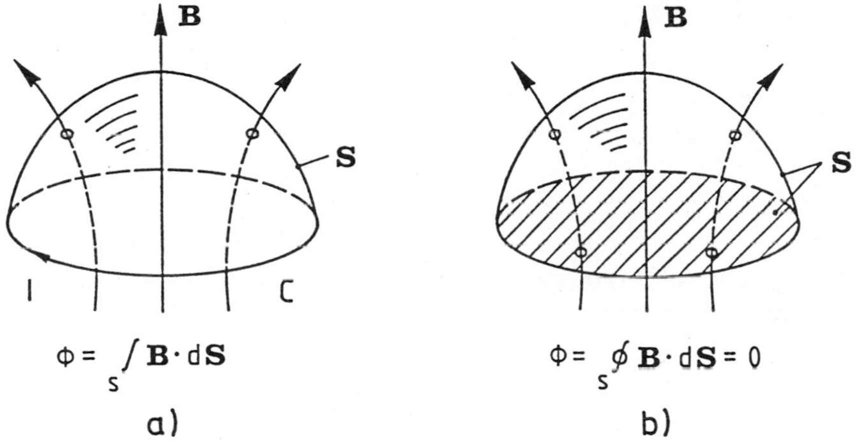

Figure 3.6: Gauss's Law of the Magnetic Field
a) open surface, b) closed surface.

Considering an arbitrary closed surface, for example the curved surface of Figure 3.6b, together with its dashed bottom, the integral equals zero for all possible magnetic vector fields. In other words, the magnetic flux leaving a closed surface equals the flux entering the surface. Because of the continuity of magnetic flux, flux-density lines do not possess starting or terminating points; they are solenoidal. Since there exist no magnetic *monopoles* (magnetic charges) but dipoles only, magnetic sources, in the sense of Chapter 2, do not exist. Magnetic fields are always vortex fields. *The source strength of magnetic fields equals zero for any field.*

Finally, it should be mentioned that abundant experimental and theoretical attempts have been made to find evidence for the existence of magnetic monopoles (encouraged by the fact that, occasionally, magnetic monopoles are a convenient fiction). However, until now, none of these attempts has proved successful.

3.2 Law of Continuity in Integral Form *Source Strength* of Current-Density Fields

Connecting a dc source to a capacitor causes in its leads a transient conduction current

$$I_c = \int \mathbf{J}_c \cdot d\mathbf{S} \tag{3-16}$$

to flow.

Between the electrodes the conduction current turns into a displacement current, Fig. 3.7

$$I_d = \int \mathbf{J}_d \cdot d\mathbf{S} \quad . \tag{3-17}$$

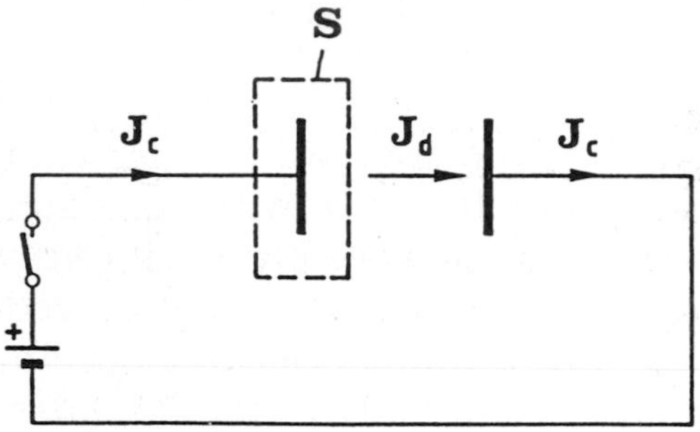

Figure 3.7: Illustration of the law of continuity for electric currents (recall Figure 3.3).

If we surround one electrode by a *closed* surface and evaluate the *source strength* of the total-current density $\mathbf{J} = \mathbf{J}_c + \mathbf{J}_d$ we obtain

$$\boxed{\oint_S \mathbf{J} \cdot d\mathbf{S} = \oint_S (\mathbf{J}_c + \mathbf{J}_d) \cdot d\mathbf{S} = 0} \,. \tag{3-18}$$

This is the *integral form of the law of continuity for total-current-density fields*. It indicates that the sum of conduction - and displacement currents entering a closed surface equals the sum of conduction- and displacement currents leaving that surface. In other words, for a closed surface this integral always vanishes. If the closed surface contains a boundary between a conductor and a dielectric (see Figure 3.3 or 3.7) the conduction current must turn into a displacement current at that boundary. Current-density lines of the total current lack starting and terminating points; that is, they are solenoidal as are the **B**-lines of magnetic fields. The vector field of the total current is a vortex field.

If we imagine the closed surface to exist in a conductive medium, for example in a busbar where the displacement current can always be neglected over the conduction current ($|\mathbf{J}_d| << |\mathbf{J}_c|$), we find

$$\boxed{\oint_S \mathbf{J}_c \cdot d\mathbf{S} = 0} \quad . \tag{3-19}$$

This is the special law of continuity for problems involving exclusively conduction current densities. It represents the *source strength of conduction fields.*

If we further imagine the closed surface to exist in a dielectric, for example in the vicinity of an antenna where the conduction current can always be neglected over the displacement current ($|\mathbf{J}_c| << |\mathbf{J}_d|$), we find

$$\boxed{\oint_S \mathbf{J}_d \cdot d\mathbf{S} = 0} \quad . \tag{3-20}$$

This is the special law of continuity for problems involving exclusively displacement currents (displacement fields $\mathbf{J}_d(\mathbf{r})$).

Frequently, the law of continuity for the total current is encountered with the displacement current density $\mathbf{J}_d$ being replaced by the time rate-of-change of the charge contained in the respective volume, i.e. dQ_V/dt. In order to obtain this form we transpose the displacement component to the equation's right side and put the differentiation with respect to time in front of the integral; thus we get

$$\oint_S \mathbf{J}_c \cdot d\mathbf{S} = -\oint_S \mathbf{J}_d \cdot d\mathbf{S} = -\oint_S \frac{d\mathbf{D}}{dt} \cdot d\mathbf{S} = -\frac{d}{dt}\oint_S \mathbf{D} \cdot d\mathbf{S} \ . \quad (3\text{-}21)$$

With $\oint_S \mathbf{D} \cdot d\mathbf{S} = Q_V$ we finally obtain

$$\boxed{\oint_S \mathbf{J}_c \cdot d\mathbf{S} = -\frac{dQ_V}{dt}} \ . \quad (3\text{-}22)$$

This is the integral form of the *law of conservation of charge.* It claims that a variation of the net charge contained in a certain volume necessarily requires a conduction current I_c to flow in to (or out of) the associated surface, allowing the amount of charge within the volume to change (needless to say that a conduction current is by definition equal to dQ/dt). Physically the conservation of charge is demonstrated from the observation that charges always appear or disappear pairwise, e.g. ionization of atoms, electrostatic charge generation, discharge of a capacitor or battery, etc.

Apparently, the law of continuity in integral form can be expressed either in terms of a time-varying electric charge or in terms of a displacement current I_d (being a measure for a time varying electric flux ψ).

In general,

$$I_d = \frac{d\psi}{dt} \quad \text{or} \quad \mathbf{J}_d = \frac{d\mathbf{D}}{dt} \quad . \tag{3-23}$$

Hence, the time-varying flux emanating from a time-varying charge and the associated displacement current are merely two different descriptive concepts for one and the same physical phenomenon. The law of conservation of charge ignores the existence of displacement currents. This is not surprising, because the displacement current concept was introduced later-on by Maxwell.

To summarize, the left sides of the second row of Maxwell's equations and the previous analogous equations of continuity possess great formal analogy and represent a particular type of integral that is also encountered in many other disciplines. This type of integral is generically called the *source-strength*. Net flux through a closed surface, total outward flux, or divergence are also commonly used, although the latter term should be reserved for the *differential* form of Gauss's law.

For any vector field **X**:

$$\boxed{\oint_S \mathbf{X} \cdot d\mathbf{S} = \begin{cases} \textit{source strength} \\ \text{net flux through} \\ \text{a closed surface} \end{cases}} \quad . \tag{3-24}$$

The notions on the relation's right side are synonymous.

$\oint_S \mathbf{X} \cdot d\mathbf{S} = 0$: The closed surface **S** is source-free, or the sum of all charges enclosed by the surface equals zero.

$\oint_S \mathbf{X} \cdot d\mathbf{S} \neq 0$: The closed surface **S** contains sources (charges in case of an electric field).

As a corollary,

$$\oint_S \mathbf{X} \cdot d\mathbf{S} = 0 \tag{3-25}$$

in a limited region does not indicate that we are dealing with a vortex field, only that the region being considered does not contain sources. Because field lines of vortex fields are solenoidal, their source strength always equals zero. The flux leaving a closed surface always equals the flux entering it.

Finally, Figure 3.8 illustrates Gauss´s law for a magnetic dipole, a positive charge, and a circuit with dielectric gap (capacitor).

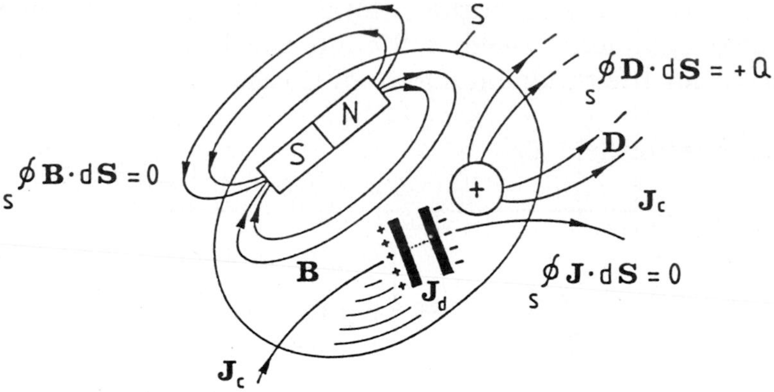

Figure 3.8: Illustration of Gauss´s laws for electric, magnetic, and conduction fields.

If a problem implies *eo ipso* a pure vortex field $\mathbf{X}_V(\mathbf{r})$, we always have

$$\oint_S \mathbf{X}_V(\mathbf{r}) \cdot d\mathbf{S} = 0. \tag{3-26}$$

3.3 Differential Form of Maxwell's Equations

Maxwell's equations in integral form make an integral or global statement about the properties of a field region which depends on the field as well as on the integration path chosen. The vortex strength of a contour C may yield a higher value the wider the contour chosen, if more vortices (of same orientation) are enclosed. In much the same way, a source strength will yield a higher value the greater the volume of the closed surface (as long as more charges of the same polarity are enclosed). Dividing a vortex strength by the area considered, and a source strength by the volume considered, yields *densities* permitting exclusive statements about the respective fields. Letting the surfaces or volumes tend towards zero yields local statements about individual field points, i.e. Maxwell's equations in differential or *point* form, Table 3.2.

DIFFERENTIAL FORM	
$\text{curl}\,\mathbf{E} = -\dfrac{\partial \mathbf{B}}{\partial t}$	$\text{curl}\,\mathbf{H} = \mathbf{J}$
$\text{div}\,\mathbf{D} = \rho$	$\text{div}\,\mathbf{B} = 0$

Table 3.2: Maxwell's equations in differential form (point form).

Looking at this system of equations without assistance is likely to turn many a student's resignation into fatalism. Eventually, however, the reader will be highly pleased, because converting Maxwell's equations from their integral form into differential form means nothing else but converting the global quantities *vortex* and *source strength* into local quantities, so-called *vortex* and *source densities*, describing discrete field points.

3.3.1 Faraday´s Induction Law in Differential Form *Vortex Density* of Electric Vortex-Fields

A convenient starting point is the integral form of Faraday´s law for matter at rest

$$\oint_C \mathbf{E} \cdot d\mathbf{r} = -\frac{d\phi}{dt} = -\int_S \frac{\partial \mathbf{B}}{\partial t} \cdot d\mathbf{S} \quad . \qquad (3\text{-}27)$$

For a certain contour C this equation yields the value of the respective vortex strength. If one wants to determine the vortex strength at a particular *field point* by allowing the contoured area to shrink to zero, the respective vortex strength, unfortunately, tends towards zero as well. To avoid this difficulty, one determines the ratio of vortex strength to the area considered. In the limit this ratio remains finite. This limit is apparently a characteristic property of the particular field point. To obtain this limit we first calculate the vortex strength for the contour C of an incremental area element ΔS whose unit normal vector $\mathbf{n}_{\Delta S}$ points in the flow direction (for this orientation the vortex strength attains its maximum value),

$$\oint_C \mathbf{E} \cdot d\mathbf{r} = -\int_{\Delta S} \frac{\partial \mathbf{B}}{\partial t} \cdot d\mathbf{S} \quad . \qquad (3\text{-}28)$$

Relating this vortex strength to the contoured area $\Delta \mathbf{S} = \Delta S \mathbf{n}_{\Delta S}$ and merely dividing by the area´s magnitude (for the same reasons as in Chapter 1) yields

$$\mathbf{n}_{\Delta S} \frac{\oint_C \mathbf{E} \cdot d\mathbf{r}}{\Delta S} = \mathbf{n}_{\Delta S} \frac{-\int_{\Delta S} \frac{\partial \mathbf{B}}{\partial t} \cdot d\mathbf{S}}{\Delta S} = \mathbf{n}_{\Delta S} \frac{-\frac{\partial}{\partial t} \int_{\Delta S} \mathbf{B} \cdot d\mathbf{S}}{\Delta S} . \quad (3\text{-}29)$$

For $\Delta S \to 0$ we obtain

$$\lim_{\Delta S \to 0} \mathbf{n}_{\Delta S} \frac{\oint_C \mathbf{E} \cdot d\mathbf{r}}{\Delta S} = \lim_{\Delta S \to 0} \mathbf{n}_{\Delta S} \frac{-\frac{\partial}{\partial t} \int_{\Delta S} \mathbf{B} \cdot d\mathbf{S}}{\Delta S} = \lim_{\Delta S \to 0} \mathbf{n}_{\Delta S} \frac{-\frac{\partial}{\partial t} \Delta\phi}{\Delta S}$$

$$= \lim_{\Delta S \to 0} \frac{\partial}{\partial t} \mathbf{n}_{\Delta S} \frac{-\Delta\phi}{\Delta S} = -\frac{\partial \mathbf{B}}{\partial t} , \quad (3\text{-}30)$$

or, more familiar,

$$\boxed{\operatorname{curl} \mathbf{E} = -\frac{\partial \mathbf{B}}{\partial t}} . \quad (3\text{-}31)$$

In each field point the vortex density curl **E** of the electric field equals the negative time rate-of-change of the local magnetic flux density. The vortex density at a discrete field point $P(\mathbf{r}_\nu)$ is a vector $\mathbf{W}(\mathbf{r}_\nu)$. The vortex-density vectors of all field points form a vector field $\mathbf{W}(\mathbf{r})$.

Calculating the vortex density for an area element $\Delta \mathbf{S}_\nu$ with orientation other than the flow direction yields the vortex density vector´s coordinate (curl $\mathbf{E}$)$_\nu$ pointing in the direction of the unit normal vector $\mathbf{n}_\nu$ of the area element.

The reader may agree with all this, but will not be highly pleased. This results from the fact that the definition of the vortex density given above is independent of a particular coordinate system and that the question how the transition towards the limit is realized in practice has not been answered yet. The reader will be relieved to find that the transition merely means partial differentiation of the electric field strength according to simple rules applicable in the respective coordinate system. For instance, in a Cartesian coordinate system the components of the vector $\mathbf{W}$ are given by

$$W_x = (\text{curl}\ \mathbf{E})_x = \frac{\partial E_z}{\partial y} - \frac{\partial E_y}{\partial z}$$

$$W_y = (\text{curl}\ \mathbf{E})_y = \frac{\partial E_x}{\partial z} - \frac{\partial E_z}{\partial x}$$

$$W_z = (\text{curl}\ \mathbf{E})_z = \frac{\partial E_y}{\partial x} - \frac{\partial E_x}{\partial y} \quad . \qquad (3\text{-}32)$$

Hence, the vortex density is

$$\mathbf{W} = \text{curl}\ \mathbf{E} = (\text{curl}\ \mathbf{E})_x\, \mathbf{a}_x + (\text{curl}\ \mathbf{E})_y\, \mathbf{a}_y + (\text{curl}\ \mathbf{E})_z\, \mathbf{a}_z \qquad (3\text{-}33)$$

or

$$\mathbf{W} = \text{curl}\ \mathbf{E} = \left(\frac{\partial E_z}{\partial y} - \frac{\partial E_y}{\partial z}\right)\mathbf{a}_x + \left(\frac{\partial E_x}{\partial z} - \frac{\partial E_z}{\partial x}\right)\mathbf{a}_y + \left(\frac{\partial E_y}{\partial x} - \frac{\partial E_x}{\partial y}\right)\mathbf{a}_z . \qquad (3\text{-}34)$$

The following simple example may serve to illustrate these equations.

Given an electric field (its units V/m are deleted for clarity)

$$\mathbf{E}(x,y,z) = E_x\,\mathbf{a}_x + E_y\,\mathbf{a}_y + E_z\,\mathbf{a}_z \quad ,$$

$$\mathbf{E}(x,y,z) = 2x^2\,\mathbf{a}_x + y^2\,x\mathbf{a}_y + 2zy\mathbf{a}_z \quad , \qquad (3\text{-}35)$$

then differentiation of its components according to the rules above readily yields

$$\text{curl}\,\mathbf{E}(x,y,z) = 2z\mathbf{a}_x + y^2\mathbf{a}_z \quad . \qquad (3\text{-}36)$$

The vortex density at a particular field point $P(x_v,y_v,z_v)$ is obtained upon insertion of its coordinates.

Alternative notation:

The vortex density of a vector field can also be written as a determinant

$$\text{curl}\,\mathbf{E}(x,y,z) = \begin{vmatrix} \mathbf{a}_x & \mathbf{a}_y & \mathbf{a}_z \\ \frac{\partial}{\partial x} & \frac{\partial}{\partial y} & \frac{\partial}{\partial z} \\ E_x & E_y & E_z \end{vmatrix} \qquad (3\text{-}37)$$

Evaluating the determinant yields the vortex density in terms of its components as given previously.

Frequently, the "del" notation is used:

$$\text{curl}\,\mathbf{E} = \nabla \times \mathbf{E} \qquad (\nabla\text{: "del", or "nabla"}). \qquad (3\text{-}38)$$

That is, the vortex density is calculated from the *cross product* of the vector del (differential-operator)

$$\nabla = \mathbf{a}_x \frac{\partial}{\partial x} + \mathbf{a}_y \frac{\partial}{\partial y} + \mathbf{a}_z \frac{\partial}{\partial z} \qquad (3\text{-}39)$$

and the vector field considered, yielding again the component representation given above.

Other coordinate systems call for similar rules (see A3).

3.3.2 Ampere´s Circuital Law in Differential Form *Vortex Density* of Magnetic Vortex-Fields

We begin with the integral form of Ampere´s law

$$\oint_C \mathbf{H} \cdot d\mathbf{r} = I = \int_S \mathbf{J} \cdot d\mathbf{S} \quad . \qquad (3\text{-}40)$$

Using the same reasoning as applied to Faraday´s induction law in the previous section, we take the limit of the ratio of vortex strength to contoured area and obtain:

$$\text{curl}\,\mathbf{H} = \lim_{\Delta S \to 0} \mathbf{n}_{\Delta S} \frac{\oint \mathbf{H} \cdot d\mathbf{r}}{\Delta S} = \lim_{\Delta S \to 0} \mathbf{n}_{\Delta S} \frac{\int_{\Delta S} \mathbf{J} \cdot d\mathbf{S}}{\Delta S} = \mathbf{J} \quad ,$$

$$\boxed{\text{curl}\,\mathbf{H} = \mathbf{J}} \quad . \qquad (3\text{-}41)$$

At each field point, the *vortex density* equals the local total current density. In formal analogy with the electric field the component representation yields

$$\text{curl}\ \mathbf{H} = \left(\frac{\partial H_z}{\partial y} - \frac{\partial H_y}{\partial z}\right)\mathbf{a}_x + \left(\frac{\partial H_x}{\partial z} - \frac{\partial H_z}{\partial x}\right)\mathbf{a}_y + \left(\frac{\partial H_y}{\partial x} - \frac{\partial H_x}{\partial y}\right)\mathbf{a}_z\ . \qquad (3\text{-}42)$$

Moreover, all other statements made at the end of the previous section apply in much the same way to the vortex density of the magnetic field.

To summarize, it can be stated that the left sides of both partial differential equations for the vortex densities of electric and magnetic vortex fields possess great formal similarity and that these definitions are encountered in similar form in many other disciplines as well. For a general vector field **X**:

$$\text{curl}\ \mathbf{X} = \nabla \times \mathbf{X} \begin{cases} \textit{vortex density} \\ \text{curl} \\ \text{vorticity} \\ \text{rot(European, except GB)} \end{cases} \qquad (3\text{-}43)$$

Let the function **X** (x,y,z) describe a vector field and let us take its curl. If the result is identically zero throughout space the vector field **X** is vortex-free, in other words, **X** is a source field. The statement curl **X** $(x_v, y_v, z_v) = 0$ indicates that the *particular* field point $P(x_v, y_v, z_v)$ is vortex-free, Figure 3.9.

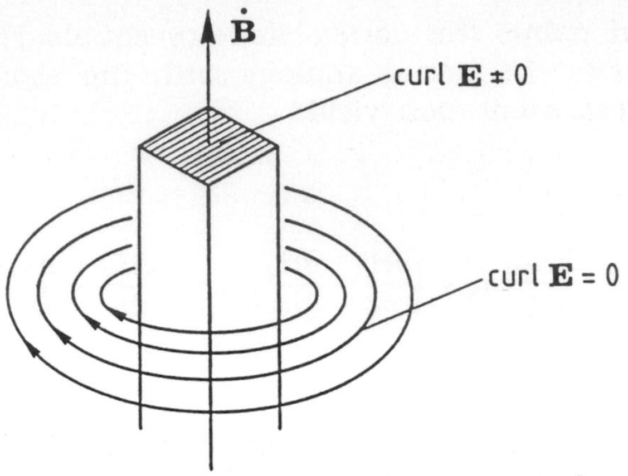

Figure 3.9: Illustration of the attribute "vortex-free"; outside the iron core the electric vortex field is *vortex- and source-free.*

Moreover, curl $\mathbf{X} = 0$ at a particular field point $P(x_V, y_V, z_V)$ or in a *limited* region does not necessarily indicate that we have a source field instead of a vortex field, merely that the region considered does not contain vortices. If a problem implies *eo ipso* a pure source field $\mathbf{X}_S(\mathbf{r})$ we always have curl $\mathbf{X}_S(\mathbf{r}) = 0$.

3.3.3 Divergence of the Electric Field *Source Density* of Electric Fields

We begin with the integral form of Gauss´s law for electric fields

$$\oint_S \mathbf{D} \cdot d\mathbf{S} = Q \quad . \tag{3-44}$$

This equation claims that the net electric flux Ψ leaving or entering a closed surface equals the charge contained in the volume considered. The integral´s value is a measure for the source strength of an electric field. One may want to determine the source strength of a field at a particular field point by allowing the volume to shrink to zero, unfortunately, the contained flux also will tend towards zero. To avoid this difficulty, one must consider the ratio of flux to volume, whose limit remains finite. This limit is apparently a characteristic property of the field point in question. To obtain this limit, we first calculate the net flux out of a small volume ΔV and relate this flux to the volume. The limit of this ratio, as ΔV approaches zero, is called *source density or divergence* of the field at that particular point

$$\operatorname{div} \mathbf{D} = \lim_{\Delta S \to 0} \frac{\oint_{\Delta S} \mathbf{D} \cdot d\mathbf{S}}{\Delta V} = \lim_{\Delta S \to 0} \frac{Q}{\Delta V} = \rho \quad ,$$

or

$$\boxed{\operatorname{div} \mathbf{D} = \rho} \quad . \tag{3-45}$$

The source density div **D** at any point of an electric field equals the local charge density. The source density of a discrete field point $P(\mathbf{r}_V)$ is a scalar, $\rho_V = \operatorname{div} \mathbf{D}_V$. The source densities of all field points form a scalar field $\rho(\mathbf{r})$. As with the curl, this transition is accomplished in practice by simple partial differentiation of the electric flux density according to simple rules applicable in the respective coordinate systems.

In a Cartesian coordinate system we have

$$\operatorname{div} \mathbf{D} = \frac{\partial \mathbf{D}}{\partial x} \mathbf{a}_x + \frac{\partial \mathbf{D}}{\partial y} \mathbf{a}_y + \frac{\partial \mathbf{D}}{\partial z} \mathbf{a}_z \tag{3-46}$$

or

$$\operatorname{div} \mathbf{D} = \frac{\partial D_x}{\partial x} + \frac{\partial D_y}{\partial y} + \frac{\partial D_z}{\partial z} = \rho\,(x,y,z) \quad . \tag{3-47}$$

The following example may serve as an illustration:

$$\mathbf{D}(x,y,z) = D_x \mathbf{a}_x + D_y \mathbf{a}_y + D_z \mathbf{a}_z = 0\mathbf{a}_x + xy\mathbf{a}_y - xz\mathbf{a}_z$$

$$\operatorname{div} \mathbf{D} = x - x = 0 \quad .$$

Hence, $\rho = 0$.

Alternative notation for divergence:

$$\operatorname{div} \mathbf{D} = \nabla \cdot \mathbf{D} \qquad (\nabla\text{: "del", or "Nabla"}) \quad . \tag{3-48}$$

The divergence is calculated from the *scalar product* of the vector (differential operator) del

$$\nabla = \mathbf{a}_x \frac{\partial}{\partial x} + \mathbf{a}_y \frac{\partial}{\partial y} + \mathbf{a}_z \frac{\partial}{\partial z} \tag{3-49}$$

and the vector field considered.

Other coordinate systems call for similar rules (see A 3).

From a given charge density and volume the associated charge can be calculated any time by means of a volume integral,

$$\boxed{Q = \int_V \rho \, dV} \tag{3-50}$$

In this instant, it can be readily understood why point charges are merely a convenient fiction. Rigorously, due to V=0, only a charge density can be associated with a point.

3.3.4 Divergence of Magnetic Fields
Source Density of Magnetic Fields

We begin with the integral form of Gauss´s law for the magnetic field

$$\oint_S \mathbf{B} \cdot d\mathbf{S} = 0 \quad . \tag{4-51}$$

This equation claims that in a magnetic field equally many flux lines leave as enter a closed surface, i.e. that the net flux through a closed surface is always zero. The closed surface does not contain sources of a magnetic field in terms of Chapter 2 (see also 3.1.4). Magnetic field lines are solenoidal, in other words, there exist no monopolar magnetic charges from which field lines in all directions would radially expand.

If one intends to demonstrate the absence of a source at a particular point of a magnetic field, allowing the contoured volume to shrink to zero, the same difficulties arise as in the previous section. Therefore, we form again the ratio of flux to volume and determine its limit as ΔV approaches zero. Then we obtain for the *source density* or *divergence* of the magnetic field

$$\operatorname{div} \mathbf{B} = \lim_{\Delta V \to 0} \frac{\oint_{\Delta S} \mathbf{B} \cdot d\mathbf{S}}{\Delta V} = \lim_{\Delta V \to 0} \frac{0}{\Delta V} = 0 \quad ,$$

$$\boxed{\operatorname{div} \mathbf{B} = 0} \quad . \tag{3-52}$$

In a Cartesian coordinate system we have

$$\operatorname{div} \mathbf{B} = \frac{\partial B_x}{\partial x} + \frac{\partial B_y}{\partial y} + \frac{\partial B_z}{\partial z} = 0 \quad . \tag{3-53}$$

The divergence div **B** of a magnetic field always equals zero, i.e. magnetic fields are source-free; magnetic field lines are solenoidal.

Moreover, all other statements made at the end of the previous section apply in much the same way.

3.4 Law of Continuity in Differential Form *Source Density* of Current-Density Fields

We start with the integral form of the law of continuity for the total current (3.2)

$$\oint_S (\mathbf{J}_c + \mathbf{J}_d) \cdot d\mathbf{S} = 0 \qquad (3\text{-}54)$$

It indicates that the *source strength* of the total-current density is identically zero. If one attempts to demonstrate the absence of a source at a particular point, allowing the contoured volume to shrink to zero, the same difficulties arise as in Sections 3.3.3 and 3.3.4. Therefore we form again the ratio flux to volume and determine its limit as ΔV approaches zero. Then we obtain the *source density* or divergence at a point

$$\text{div}\,\mathbf{J} = \text{div}\,(\mathbf{J}_c + \mathbf{J}_d) = \lim_{\Delta V \to 0} \frac{\oint_S (\mathbf{J}_c + \mathbf{J}_d) \cdot d\mathbf{S}}{\Delta V} = 0 \quad ,$$

$$\boxed{\text{div}\,\mathbf{J} = \text{div}\,(\mathbf{J}_c + \mathbf{J}_d) = 0} \quad . \qquad (3\text{-}55)$$

This is the *differential form of the law of continuity of the total current.*

Expressed in a Cartesian coordinate system we have

$$\operatorname{div} \mathbf{J} = \frac{\partial J_x}{\partial x} + \frac{\partial J_y}{\partial y} + \frac{\partial J_z}{\partial z} = 0 \quad . \tag{3-56}$$

The divergence of the total-current field is always zero; the total-current field is source-free in the sense of Chapter 2. On boundaries of dielectrics, conduction-current-density lines become displacement-current-density lines. Current-density lines of the total current are always solenoidal (recall Figure 3.3 and 3.7)

If we imagine a field point inside a conductor, for example in a busbar where the displacement current can always be neglected over the conduction current ($|\mathbf{J}_d| \ll |\mathbf{J}_c|$), we find

$$\boxed{\operatorname{div} \mathbf{J}_c = 0} \quad . \tag{3-57}$$

This is the special law of continuity for problems involving exclusively conduction current densities. It represents the *source density* of pure conduction fields $\mathbf{J}_c(\mathbf{r})$.

If we further imagine a field point in the vicinity of an antenna where the conduction current can always be neglected over the displacement current ($|\mathbf{J}_c| \ll |\mathbf{J}_d|$), we find

$$\boxed{\operatorname{div} \mathbf{J}_d = 0} \quad . \tag{3-58}$$

This is the special law of continuity for problems involving exclusively displacement currents. It represents the *source density* of pure displacement fields $\mathbf{J}_d(\mathbf{r})$.

Frequently, the differential form of the law of continuity for the total current is encountered with the displacement current density replaced by the time rate-of-change of the charge density at a particular point, i.e. $d\rho/dt$.

In order to obtain this form we transpose the displacement component to the equation´s right side and put the differentiation with respect to time in front of the differential operator div,

$$\operatorname{div} \mathbf{J}_c = - \operatorname{div} \mathbf{J}_d = - \operatorname{div} \frac{\partial \mathbf{D}}{\partial t} = - \frac{\partial}{\partial t} \operatorname{div} \mathbf{D} \quad . \qquad (3\text{-}59)$$

With div $\mathbf{D} = \rho$ we finally obtain

$$\boxed{\operatorname{div} \mathbf{J}_c = - \frac{\partial \rho}{\partial t}} \quad . \qquad (3\text{-}60)$$

This is the differential form of the *law of conservation of charge.* It relates the *source density* of the conduction current density to the variation of the charge density at a particular field point.

We note that the differential form of the law of continuity of the total-current density can be expressed in terms of a time-varying charge density as well as in terms of the displacement current density $\mathbf{J}_d$ (defined as the time rate-of-change of the electric flux-density).

In general,

$$I_d = \frac{d\psi}{dt} \quad \text{or} \quad \mathbf{J}_d = \frac{d\mathbf{D}}{dt} \quad . \qquad (3\text{-}61)$$

Hence, the time-varying electric flux density originating from a variation of charge density and the displacement current density $\mathbf{J}_d$ are merely two different descriptive concepts for one and the same physical phenomenon.

The law of conservation of charge ignores the existence of displacement currents. This is not surprising, because the displacement current concept was introduced later-on by Maxwell.

To summarize, it can be stated that the left sides of the differential equations for the source densities of electric and magnetic fields, and the left sides of the previous analogous equations for current fields possess great formal similarity and that these relations, although of different physical quality, are encountered in many other disciplines. For a general vector field $\mathbf{X}$ we have

$$\boxed{\operatorname{div}\mathbf{X} = \nabla\cdot\mathbf{X} \quad \begin{cases} \textit{source density} \\ \text{divergence} \end{cases}} \quad . \tag{3-62}$$

Let the function $\mathbf{X}$ (x,y,z) describe a vector field and let us take its divergence. If the result is identically zero the vector field $\mathbf{X}$ is source-free, i.e. we are dealing with a vortex field. The statement div $\mathbf{X}$ $(x_v,y_v,z_v) = 0$ indicates that there is no source at the particular field point $P(x_v,y_v,z_v)$. For instance, within the dielectric of a parallel-plate capacitor we have everywhere div $\mathbf{D}$ = 0, though we are dealing with a source field.

Thus, the fact that div $\mathbf{X}$ = 0 at a particular field point or in a *limited* region does not necessarily indicate that we have a vortex field instead of a source field, merely that the region considered is free of sources.

If a particular problem implies *eo ipso* a pure vortex field $\mathbf{X}_v$ $(\mathbf{r})$ we always have div $\mathbf{X}_v(\mathbf{r}) = 0$.

The concepts of the first four sections of this chapter summarize in terms of Maxwell's equations according to Table 3.3.

SOURCE FIELDS	SOURCE *STRENGTH*	$\oint_S \mathbf{D}\cdot d\mathbf{S} = \int_V \rho dV$	$\oint_S \mathbf{B}\cdot d\mathbf{S} = 0$
	SOURCE *DENSITY*	$\nabla\cdot\mathbf{D} = \rho$	$\nabla\cdot\mathbf{B} = 0$
VORTEX FIELDS	VORTEX *STRENGTH*	$\oint_C \mathbf{E}\cdot d\mathbf{r} = \int_S \frac{\partial\mathbf{B}}{\partial t}\cdot d\mathbf{S}$	$\oint_C \mathbf{H}\cdot d\mathbf{r} = \int_S \mathbf{J}\cdot d\mathbf{S}$
	VORTEX *DENSITY*	$\nabla\times\mathbf{E} = -\dot{\mathbf{B}}$	$\nabla\times\mathbf{H} = \mathbf{J}$

Table 3.3: Comparative survey of the paired concepts *source strength / source density*, and *vortex strength / vortex density* for electric and magnetic fields.

3.5 Maxwell's Equations in Complex Notation (Phasor Form)

In time-varying fields the field vectors are not only functions of position but also of time, e.g. $\mathbf{E}$ (x,y,z,t) or $\mathbf{E}$ ($\mathbf{r}$,t). Hence, the partial differential equations exhibit both space and time variables.

To simplify the solution of these differential equations, one frequently limits oneself to harmonic variations (sinusoidal wave forms). Then, by means of complex notation, the time dependence can be eliminated because the time factor $e^{j\omega t}$ cancels (see A 5). Thus, the field quantities become complex amplitudes depending on spatial variables only. Instead of partially differentiating the quantities with respect to time, they are multiplied by $j\omega$. Hence, Maxwell's equations in differential form and in complex notation are in Table 3.4.

COMPLEX NOTATION OF MAXWELL'S EQUATIONS IN DIFFERENTIAL FORM	
curl $\underline{\mathbf{E}} = -j\omega\underline{\mathbf{B}}$	curl $\underline{\mathbf{H}} = \underline{\mathbf{J}} = (\sigma + j\omega\varepsilon)\,\underline{\mathbf{E}}$
div $\underline{\mathbf{D}} = \rho$	div $\underline{\mathbf{B}} = 0$

Table 3.4: Complex notation of Maxwell's Equations.

3.6 Integral Theorems of Stokes and Gauss

The previous sections have attempted to illustrate the physical essence of Maxwell's equations in integral and differential form, employing the concepts of *vortex strength* versus *vortex density* and *source strength* versus *source density*. Of course, these concepts are based on well founded mathematical principles. In keeping with the philosophy of this book, a brief discussion of the integral theorems of Stokes and Gauss is now appropriate. These allow a transfer of Maxwell's equations from their integral form into their differential form, in other words to transfer vortex strengths into vortex densities and source strengths into source-densities, and vice versa.

STOKES' THEOREM:

Stokes' theorem links *vortex strength* and *vortex density* of a vector field

$$\boxed{\oint_C \mathbf{X}\cdot d\mathbf{r} = \int_S \text{curl}\,\mathbf{X}\cdot d\mathbf{S}} \quad . \tag{3-63}$$

A vortex strength equals the surface integral over the respective vortex density, the integration undoing the formation of the ratio of vortex-strength to the differential area dS. In other words, the right side restores the vortex strength. For a uniform field, the integral on the right side simplifies to the multiplication of a vortex density by an area. For instance, substituting for **X** the magnetic field strength **H**, one obtains from the magnetic vortex density curl **H** = **J**

$$\oint_C \mathbf{H} \cdot d\mathbf{r} = \int_S \mathrm{curl}\, \mathbf{H} \cdot d\mathbf{S} = \int_S \mathbf{J} \cdot d\mathbf{S} = I \quad , \qquad (3\text{-}64)$$

which has already been found to be valid in Chapter 3.1.2.

GAUSS´S THEOREM:

Gauss´s theorem links *source strength* and *source density* of a vector field and generically includes the statements of Gauss´s laws of the electric and magnetic field (see 3.1.3 and 3.1.4)

$$\boxed{\oint_S \mathbf{X} \cdot d\mathbf{S} = \int_V \mathrm{div}\, \mathbf{X} dV} \quad . \qquad (3\text{-}65)$$

A source strength equals the volume integral over the respective source density, the integration undoing the formation of the ratio of source strength to a differential volume dV. In other words, the right side restores the source strength (net flux). For a uniform source distribution the integral on the right side simplifies to the multiplication of the source density by a volume.

For instance, substituting for **X** the electric flux density **D**, one obtains with the electric source density div **D** = ρ

$$\oint_S \mathbf{D} \cdot d\mathbf{S} = \int_V \mathrm{div}\mathbf{D}\, dV = \int_V \rho dV = Q \qquad (3\text{-}66)$$

which has already been found to be valid in Chapter 3.3.3.

Explicit integration in Stokes'and Gauss's theorems can be readily carried out for a particular coordinate system; however, this is beyond the scope of this book.

3.7 Network Model of Magnetic Induction

According to Faraday's induction law, time-varying magnetic fields induce electric voltages in conductor loops (see 3.1.1). These voltages can be calculated

- in the time-domain (for arbitrary wave forms) from

$$v_i(t) = -\frac{d\phi}{dt} = -\dot{\phi} \quad , \tag{3-67}$$

- in the frequency domain (for sinusoidal or time-harmonic wave forms, steady-state) from

$$\underline{V}_i\,(j\omega) = -\,j\omega\underline{\phi} \quad . \tag{3-68}$$

In the latter equation, $\underline{V}$ and $\underline{\phi}$ are complex amplitudes (see A 5).

Likewise, *self-induced voltages*, i.e. voltages induced by the magnetic fields of time-varying currents in their particular circuits, can be calculated from the equations above. The induction phenomenon is of fundamental importance for the whole world of electrical engineering, therefore the task of modeling magnetic induction by an equivalent circuit is frequently encountered. Since in network theory typical field quantities, for example magnetic flux etc., are not defined, one must manage with passive components, voltages and currents and their sources.

For the calculation of voltages induced into complex conductor structures by transient electromagnetic fields, as well as for an intimate understanding of the principles of transformers and rotating machines, representation of the induction effect by a voltage source e(t) or $\underline{E}$ (jω) is an essential concept, Figure 3.10.

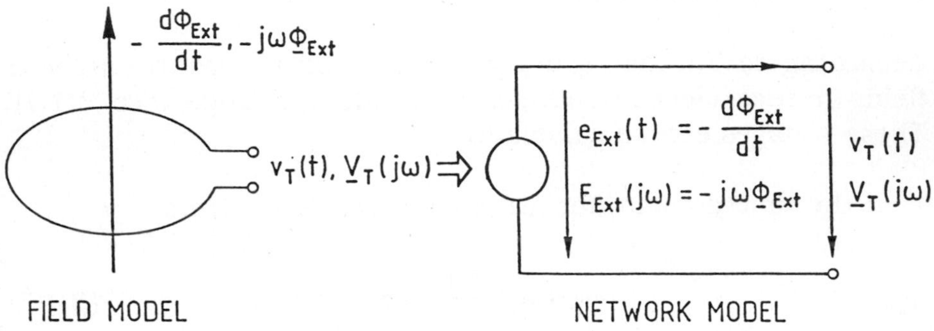

Figure 3.10: Field- and network-model of the induction effect. Representation of an induced voltage in the time and frequency domain as source voltage with direction opposite to current flow; $\underline{V}_T$ (jω), $v_T(t)$, terminal voltages.

The fact that we deal with a voltage induced by an *external* field is indicated by the subscript "Ext". Furthermore, all following definitions are given both for time and frequency domains.

Applying Kirchhoff's voltage law ΣV = 0 to an open circuit yields

$$e_{Ext}(t) - v_T(t) = 0 \quad \rightarrow \quad v_T(t) = e_{Ext}(t)$$

or

$$\underline{E}_{Ext}(j\omega) - \underline{V}_T(j\omega) = 0 \quad \rightarrow \quad \underline{V}_T(j\omega) = \underline{E}_{Ext}(j\omega) \; . \qquad (3\text{-}69)$$

Frequently, an induced voltage is represented as an *electromotive force (emf)* with the same direction as the current flow.

Employing Kirchhoff's voltage law in its original form, $\Sigma emf = \Sigma \Delta V$, yields the same results.

The equivalent circuit depicted in Figure 3.10 applies only to open circuits, i.e. unloaded systems. Under load conditions a current i(t) or $\underline{I}$ will flow whose magnetic field ϕ_I will induce an additional voltage into the circuit. The fact that the self-induced voltage results from this current's magnetic field is indicated by the subscript I. In an equivalent circuit a self-induced voltage can be represented as a source voltage (as before) or as an inductive voltage drop. The first representation is essential for the understanding of electrical machines, the latter is more common in network theory, Figure 3.11.

SELF-INDUCED SOURCE VOLTAGE

$$e_I(t) = -\frac{d\phi_I}{dt} = -L\frac{di(t)}{dt}$$

$$\underline{E}_I(j\omega) = -j\omega\underline{\phi}_I = -j\omega L\underline{I}(j\omega)$$

$$e_{Ext}(t) = -\frac{d\phi_{Ext}}{dt}$$

$$\underline{E}_{Ext}(j\omega) = -j\omega\underline{\phi}_{Ext}$$

INDUCTIVE VOLTAGE DROP

$$v_L(t) = \frac{d\phi_I}{dt} = L\frac{di(t)}{dt}$$

$$\underline{V}_L(j\omega) = j\omega\underline{\phi}_I = j\omega L\underline{I}(j\omega)$$

$$e_{Ext}(t) = -\frac{d\phi_{Ext}}{dt}$$

$$\underline{E}_{Ext}(j\omega) = -j\omega\underline{\phi}_{Ext}$$

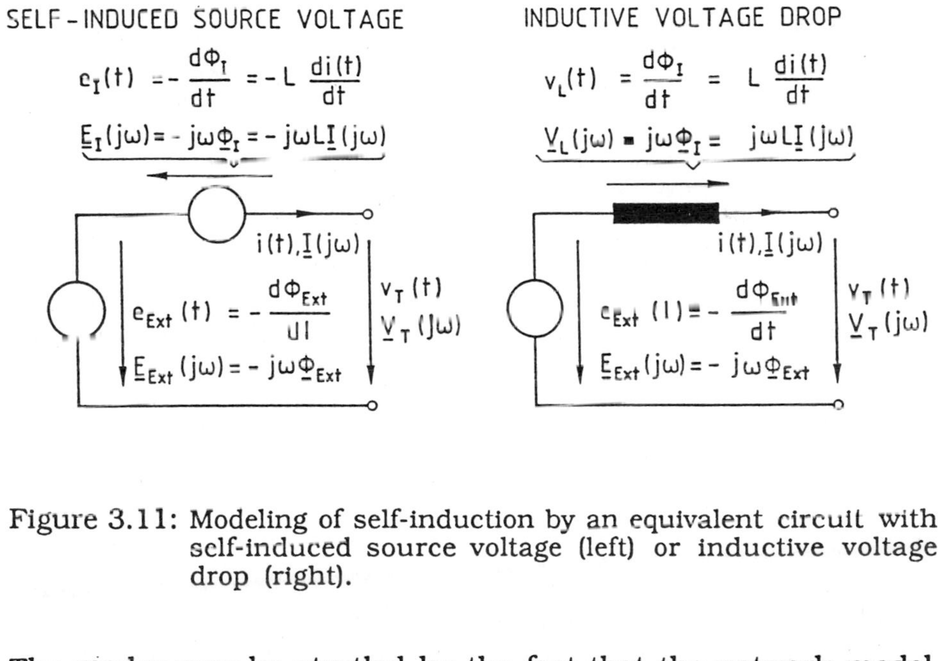

Figure 3.11: Modeling of self-induction by an equivalent circuit with self-induced source voltage (left) or inductive voltage drop (right).

The reader may be startled by the fact that the network model with self-induced source voltages shows identical directions for the induced and self-induced voltages because, according to Lenz's law, both effects should counteract each other. However, it should be noted that the arrows of the voltage source symbols define only the sign according to which the source voltages are to be summed up. The sign of the physical effect is contained in the subscript of the magnetic fluxes $\underline{\phi}_{Ext}$ and $\underline{\phi}_I$, in other words, it is latent in the complex amplitude´s argument, Figure 3.12.

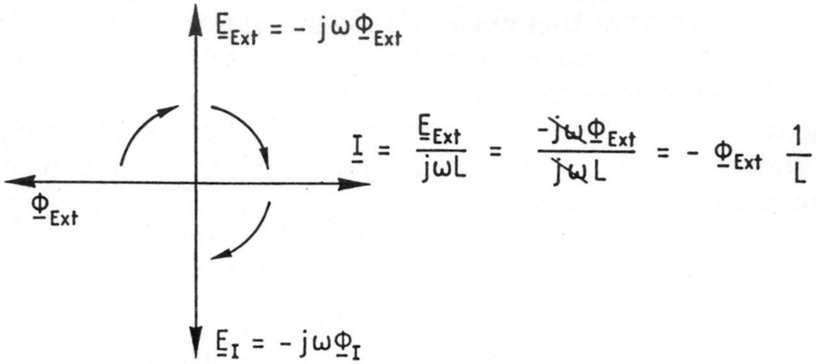

Figure 3.12: Phase shift of $\underline{E}_{Ext}$ and $\underline{E}_I$ for a load impedance Z=0 (short circuit).

A flux $\underline{\phi}_{Ext}$ induces a voltage $\underline{E}_{Ext}$ lagging 90°. This voltage drives a current $\underline{I}$ with additional 90° time-shift through the inductance L. The associated magnetic flux $\underline{\phi}_I$ being in phase with the current $\underline{I}$, induces a voltage $\underline{E}_I$ which exhibits a time-shift of 180° with respect to $\underline{E}_{Ext}$. Thus the voltage $\underline{E}_{Ext}$ driving the current $\underline{I}$ leads the flux $\underline{\phi}_I$ by 90°.

The self-induced voltage of a coil, often referred to as *counter emf*, always opposes, or compensates, the current-driving applied voltage. For instance, the magnetizing current of an unloaded transformer adjusts its waveform precisely in such a way that the self-induced voltage, induced by its flux in the primary, in every instant balances the applied voltage.

If the applied voltage varies harmonically with time, the counter emf and the flux inducing it must vary harmonically as well. However, since the magnetic conductivity of an iron core near saturation decreases with increasing flux, a sinusoidal flux can only be maintained by a current waveform possessing relatively higher instantaneous values in the peak region. This effect leads to the well known nonsinusoidal waveform of the magnetizing current of transformers and illuminates the frequently encountered statement "... the flux follows the voltage". Further, the question of whether a transformer will exhibit saturation effects or not is determined by the applied voltage, not by its secondary load. If a transformer shows no saturation under open-circuit conditions, it will never show saturation (at the same primary voltage) even when the secondary is short circuited.

In order to avoid confusion, one should either talk about

- self-induced *source voltages* with a direction opposite to that of current flow, (having Faraday´s induction law in mind) or about

- inductive *voltage drops* with the same direction as current flow, (having Ohm´s law in mind, in which case the existence of an induction law should be ignored).

For the development of descriptive models of complex systems, e.g. simulation of transient electromagnetic fields coupling into linear conductor structures or generator models for stability studies in electric energy systems, the concepts of induced and self-induced source voltage are essential.

4 Gradient, Potential, Potential Function

In physics numerous quantities are coupled via *derivatives* or *indefinite integrals*, for example path and velocity in mechanics

$$v(t) = \frac{ds(t)}{dt} \quad \text{or} \quad s(t) = \int v(t)\, dt + C \quad , \qquad (4\text{-}1)$$

or voltage and current of a constant inductance in network theory

$$u(t) = L\frac{di(t)}{dt} \quad \text{or} \quad i(t) = \frac{1}{L}\int u(t)\, dt + C \quad . \qquad (4\text{-}2)$$

In like manner, in field theory the electric field strength $\mathbf{E}(\mathbf{r})$ of a source field and the potential function $\varphi(\mathbf{r})$ of the associated charge aggregate are related to each other,

$$\mathbf{E}(\mathbf{r}) = -\frac{d\varphi(\mathbf{r})}{d\mathbf{r}} \quad \text{or} \quad \varphi(\mathbf{r}) = -\int \mathbf{E}(\mathbf{r}) \cdot d\mathbf{r} + C \quad . \qquad (4\text{-}3)$$

In the first example the integration constant C represents a previously covered distance s_o which is independent of the present time variable, in the second, a preexisting dc current I_o, and in the last, a constant potential φ_o independent of the spatial variable $\mathbf{r}$. Given an upper and a lower bound, the indefinite integrals change to definite integrals with vanishing integration constant. Then, the quantities s(t), i(t) and, $\varphi(\mathbf{r})$ represent no longer variables of a function but a particular distance Δs, a current change Δi, or a potential difference $\Delta\varphi$, respectively. Whereas in the first two examples path and current are notions of their own, *potential* and *potential functions* represent a universal generic concept which is encountered in many disciplines and which may possess many different physical interpretations. In this context, we are talking about the *electric potential.* Frequently, potential serves only as a mathematical surrogate function or artifice, via which a field strength $\mathbf{E}(\mathbf{r})$ can be evaluated more easily compared with a direct solution of an equation with $\mathbf{E}$ as dependent variable.

Electrostatic charges furnish their environment with an electric field whose existence can be demonstrated by its forces on objects brought into this region. Mathematically, this field can be modelled either by a vector field-strength function $\mathbf{E}(\mathbf{r})$ or by a scalar potential function $\varphi(\mathbf{r})$. The latter can be derived from the field-strength function via the indefinite integral

$$\int \mathbf{E}(\mathbf{r}) \cdot d\mathbf{r} \qquad (4\text{-}4)$$

to within an integration constant $C = \varphi_0$.

Hence, in absence of a spatially independent potential φ_0, a scalar potential field can be uniquely defined for each electric source field,

$$\boxed{\varphi(\mathbf{r}) = -\int \mathbf{E}(\mathbf{r}) \cdot d\mathbf{r}} \quad . \qquad (4\text{-}5)$$

The inverse of the potential definition is

$$\mathbf{E} = -\frac{d\varphi(\mathbf{r})}{d\mathbf{r}} = -\ \text{grad}\ \varphi(\mathbf{r}) \quad . \tag{4-6}$$

The derivative of φ with respect to the space variable **r**, $d\varphi(\mathbf{r})/d\mathbf{r}$, is called *gradient* of the potential field $\varphi(\mathbf{r})$. Therefore, the differential operator $d/d\mathbf{r}$ is frequently written *grad*. Because of their great practical importance the terms *gradient* and *potential* will be discussed in more detail in the following sections.

4.1 Gradient of a Scalar Field

As has already been mentioned, an electric source field can be characterized by a *vector field-strength function* $\mathbf{E}(\mathbf{r})$, or $\mathbf{E}(x,y,z)$, as well as by a *scalar mathematical surrogate function* $\varphi(\mathbf{r})$ or $\varphi(x,y,z)$, referred to as the *potential function*. In a Cartesian coordinate system, both functions are related via the following differential operation

$$\mathbf{E}(x,y,z) = -\frac{d\varphi(\mathbf{r})}{d\mathbf{r}} = -\left(\frac{\partial\varphi}{\partial x}\mathbf{a}_x + \frac{\partial\varphi}{\partial y}\mathbf{a}_y + \frac{\partial\varphi}{\partial z}\mathbf{a}_z\right) \quad . \tag{4-7}$$

To illustrate this equation, we return to our very first example of a scalar field, the temperature distribution in a living-room. If the scalar function $\varphi(x,y,z)$ were to represent the temperature distribution $T(x,y,z)$, then $\mathbf{E}(x_1,y_1,z_1)$ would indicate magnitude and direction of the maximum temperature rate-of-change at the field point $P(x_1,y_1,z_1)$, in other words, the maximum local slope of the function $\varphi(x,y,z)$ at that point. This slope is called gradient of the function $\varphi(x,y,z)$ at the field point $P(x_1,y_1,z_1)$. In our imagination the gradient may be visualized as the familiar slope of a straight line when we restrict ourselves to a one-dimensional function $\varphi(x)$. The gradient points in the direction of increasing level (in contrast, **E** points to decreasing levels!) In general, the gradient of a scalar field is

$$\boxed{\mathbf{X}(\mathbf{r}) = \text{grad}\ \varphi(\mathbf{r})} \quad . \qquad (4\text{-}8)$$

The gradient at a field point P_ν is a vector $\mathbf{X}_\nu(\mathbf{r}_\nu)$. The gradients of all points form a vector field $\mathbf{X}(\mathbf{r})$. Early in electrical science, the electric field strength was defined to point from higher to lower potentials (from plus to minus). Therefore, the *gradient* of the *electric potential* $\varphi(\mathbf{r})$ was given a negative sign,

$$\boxed{\mathbf{E}(\mathbf{r}) = -\ \text{grad}\ \varphi(\mathbf{r})} \quad . \qquad (4\text{-}9)$$

Expanded in a Cartesian coordinate system, for instance, the differential operator grad takes the following form,

$$\text{grad} = \frac{\partial}{\partial x}\,\mathbf{a}_x + \frac{\partial}{\partial y}\,\mathbf{a}_y + \frac{\partial}{\partial z}\,\mathbf{a}_z \quad . \qquad (4\text{-}10)$$

Frequently, the symbolic notation grad $\varphi = \nabla\varphi$ is used, in which the vector ∇ (read "del") represents the differential operator grad. Other coordinate systems require different expansions (see A 3).

The equation $\mathbf{E}$ = - grad φ can be derived formally as well. The electric field surrounding electrostatic charges is a pure source field. Hence, it is vortex-free, and we have curl $\mathbf{E} = 0$. Furthermore, it can be shown that for any analytical function $\varphi(x,y,z)$ of a scalar potential-field, curl grad $\varphi = 0$. In other words, differentiating $\varphi(x,y,z)$ first according to the recently given rule grad and then according to the rule curl **X** defined in Chapter 3.3.1, the result is found to be identically zero.

Equating both equations yields

$$\text{curl}\,\mathbf{E} = \text{curl grad}\,\varphi \quad . \tag{4-11}$$

Integrating this equation (that is, undoing the curl operation on both sides, $\{\text{curl}\}^{-1}$) yields **E** = grad φ. The negative sign in electrical engineering is obtained by multiplying curl grad φ = 0 by (-1) before equating. In that case one obtains

$$\boxed{\mathbf{E} = -\,\text{grad}\,\varphi} \quad . \tag{4-12}$$

The integration performed above is to be understood in that sense in which the solving of differential equations is frequently spoken of . The symbol $\{\text{curl}\}^{-1}$ is a new integral operator which is inverse to the well known differential operator curl (considering a unique solution, which is, in general, provided by boundary conditions specific to the problem). Because **E** and grad φ are *eo ipso* source fields, the result is unique (apart from a position-independent integration constant φ_o which we set equal to zero, see 4.2). This straight-forward procedure is equivalent to the usual statements of the type "... vortex-free vector fields **E**, characterized by curl **E** = 0, can, because of curl grad φ=0, be represented as the gradient of a potential φ ...". The interested reader may look up Appendix 4 to see what $\{\text{curl}\}^{-1}$ mathematically stands for.

Concluding, it is to be noted that the electric field strength **E**(x,y,z) can be obtained from a given potential function φ(x,y,z) by a simple differentiating process. However, it should not be hidden that the true work lies in finding the scalar function φ(x,y,z).

4.2 Potential and Potential Function of Electrostatic Fields

An electrostatic field contains stored potential energy which has been invested during (i) the charge generation via ionization of neutral particles, and (ii) transportation to their respective position. In order to shed more light on the concept of potential which was only formally introduced so far, we calculate this energy for an electric charge that is brought from infinity into an existing electric field. Depending on the charge's polarity and its final position, the existing field (energy storage facility) receives or releases energy, Figure 4.1.

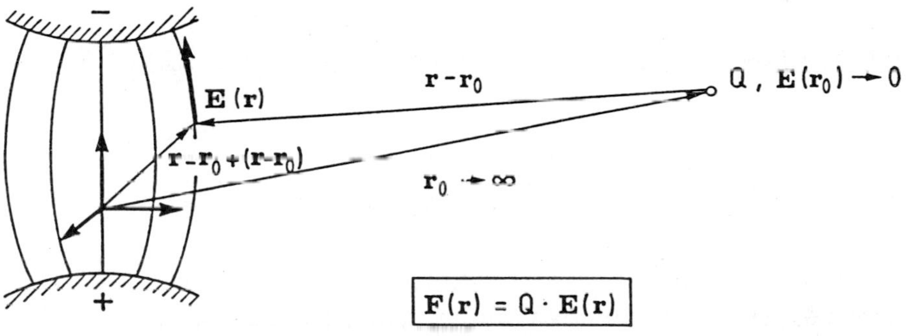

Figure 4.1: Calculation of the energy transferred during transportation of a charge from a field-free region ($\mathbf{r}_0=\infty$) into an electric field.

Let the charge Q be so small that it shall not disturb the field $\mathbf{E}(\mathbf{r})$ significantly. Then, the field's force on the charge is $\mathbf{F}(\mathbf{r}) = Q\mathbf{E}(\mathbf{r})$. With this force, the work done by the field or the energy invested by the field, can be calculated from

$$W = \int_{\mathbf{r}_0=\infty}^{\mathbf{r}} \mathbf{F}(\mathbf{r}) \cdot d\mathbf{r} = \int_{\infty}^{\mathbf{r}} Q\mathbf{E}(\mathbf{r}) \cdot d\mathbf{r} = Q \int_{\infty}^{\mathbf{r}} \mathbf{E}(\mathbf{r}) \cdot d\mathbf{r} \quad . \qquad (4\text{-}13)$$

This energy depends on both the amount and sign of the charge Q and on the field-specific component

$$\int \mathbf{E}(\mathbf{r}) \cdot d\mathbf{r} \quad , \tag{4-13}$$

which we have already encountered as potential. Next we isolate the latter by relating the energy to the transported amount of charge and by representing the electric field strength as the gradient of the scalar potential function $\varphi(\mathbf{r})$. Simultaneously, we change the sign, that is, we no longer consider the field's work but rather the energy contributed by an external actor,

$$-\frac{W}{Q} = -\int_{\infty}^{\mathbf{r}} \mathbf{E}(\mathbf{r}) \cdot d\mathbf{r} = \int_{\infty}^{\mathbf{r}} \operatorname{grad} \varphi(\mathbf{r}) \cdot d\mathbf{r} = \int_{\infty}^{\mathbf{r}} d\varphi = \varphi(\mathbf{r}) - \varphi(\infty) \quad . \tag{4-14}$$

In this equation $\varphi(\mathbf{r})$ represents the specific "potential" of an electric field at a field point $\mathbf{r}$ to do work, $\varphi(\infty)$ the specific "potential" of a field point at infinity to do work, and $-W/Q$ the difference between the two. Since the field strength at infinity is zero and transportation of a charge in a field-free region does not require work, we define $\varphi(\infty) = 0$. With this presumption the spatially dependent specific ability of a field $\mathbf{E}(\mathbf{r})$ to do work is

$$\boxed{\varphi(\mathbf{r}) = -\int \mathbf{E}(\mathbf{r}) \cdot d\mathbf{r}} \quad . \tag{4-15}$$

The specific ability to do work at a particular field point $P(\mathbf{r}_1)$, or $P(x_1, y_1, z_1)$ is called its potential $\varphi(\mathbf{r}_1)$ or $\varphi(x_1, y_1, z_1)$. The scalar function $\varphi(x,y,z)$ is called the potential function. It is a scalar field which can be constructed into a vector field $\mathbf{E}(x,y,z)$

according to the definition given above. The potential of an electrostatic field is measured in *volts*.

The potential derived under the assumption $\varphi(\infty) = 0$ is often called the *absolute* potential. The difference in ability to do work of two field points $\mathbf{r}_1$ and $\mathbf{r}_2$ is called the electric voltage V_{12} between these two points,

$$\varphi(\mathbf{r}_1)-\varphi(\mathbf{r}_2) = -\int_{\infty}^{\mathbf{r}_1} \mathbf{E}(\mathbf{r}) \cdot d\mathbf{r} + \int_{\infty}^{\mathbf{r}_2} \mathbf{E}(\mathbf{r}) \cdot d\mathbf{r} = \int_{\mathbf{r}_1}^{\mathbf{r}_2} \mathbf{E}(\mathbf{r}) \cdot d\mathbf{r} = V_{12} \,. \qquad (4\text{-}16)$$

Conversely, the potential difference between two field points $\mathbf{r}_1$ and $\mathbf{r}_2$ can be calculated from the voltage V_{12}

$$V_{12} = \int_{\mathbf{r}_1}^{\mathbf{r}_2} \mathbf{E}(\mathbf{r}) \cdot d\mathbf{r} = -\int_{\mathbf{r}_1}^{\mathbf{r}_2} \operatorname{grad} \varphi(\mathbf{r}) \cdot d\mathbf{r} = -\int_{\mathbf{r}_1}^{\mathbf{r}_2} d\varphi = -\varphi(\mathbf{r}) \Big|_{\mathbf{r}_1}^{\mathbf{r}_2}$$

$$= \varphi(\mathbf{r}_1) - \varphi(\mathbf{r}_2) \,; \qquad \text{e.g.} \quad 11\text{V} - 3\text{V} = 8\text{V} \,. \qquad (4\text{-}17)$$

Frequently, the grounded parts of a circuit are assigned zero potential (ground potential), which can be illustrated either by definition – $\varphi_\infty = 0$ – or by a fictitious conductor connecting ground and infinity. Thus, the grounded terminal of a voltage source is at ground potential, its *live terminal at high-voltage* potential.

The inverse of the definition of a potential is an equation that has already been presented at the beginning of this chapter,

$$\boxed{\mathbf{E}(\mathbf{r}) = -\,\mathrm{grad}\,\varphi(\mathbf{r})} \quad . \tag{4-18}$$

From this equation it can be recognized that the potential at a field point is not a unique quantity. For instance, given a potential function $\varphi(\mathbf{r})$ one can superpose an arbitrary spatially independent potential φ_0 = const. whose existence upon gradient operation can no longer be perceived because it does not contribute to the electric field $\mathbf{E}(\mathbf{r})$. Therefore,

$$\boxed{\mathbf{E}(\mathbf{r}) = -\,\mathrm{grad}\,\varphi(\mathbf{r}) = -\,\mathrm{grad}\,[\varphi(\mathbf{r}) + \varphi_0] := -\,\mathrm{grad}\,\varphi^*(\mathbf{r})} \quad . \tag{4-19}$$

The potential function $\varphi(\mathbf{r})$ yields precisely the same electric field as the potential function $\varphi^*(\mathbf{r})$. Only upon a reasonable convention regarding φ_0, e.g. the choice $\varphi(\infty) = 0$ in the example above, will the potential be unique. The convention $\varphi(\infty) = 0$ is convenient in many cases, although not mandatory. Depending on the specific problem, another choice may make more sense, for instance with boundary-value problems. This makes it apparent once more that the concept of potential is of mathematical rather than physical nature.

Dividing the potential function of a capacitor´s field by the voltage applied reveals that the related potential function is exclusively a property of space or its boundaries, not a function of voltage or anything electric. Thus, the related potential function of a particular electric field is identical with the normalized potential function of a gravitational field with the same boundaries.

As a corollary, the potential concept is not exclusive to the electric field but is an interdisciplinary mathematical artifice employed to characterize vortex-free fields in an alternative manner (apart from field strength) and to allow a simpler evaluation of an unknown vector field by mere differentiation of a scalar function. In general, each vortex-free region of an arbitrary vector field $\mathbf{X}(\mathbf{r})$ can be assigned a formal surrogate quantity - the scalar potential function $\varphi(\mathbf{r})$ - via

$$\boxed{\mathbf{X} = \text{grad}\ \varphi} \quad . \tag{4-20}$$

Taking the vortex density on both sides of this equation gives

$$\text{curl}\ \mathbf{X} = \text{curl grad}\ \varphi = 0 \quad , \tag{4-21}$$

since the differential operators curl and grad sequentially applied to any scalar field will always yield zero (which the reader should verify for a simple analytic function). Consequently, any function $\mathbf{X}(\mathbf{r})$ can be derived via the gradient operation from a scalar potential function $\varphi(\mathbf{r})$ which complies with the criterion of curl $\mathbf{X} = 0$ in the region considered, in other words one which is vortex-free. Finally, it should be pointed out once more that the considerations above apply only to static and quasi-static (slowly varying) fields. The latter will be discussed in more detail in Chapter 6.

4.3 Development of the Potential Function from a Given Charge Distribution

According to Gauss´s law of the electric field, $\oint \mathbf{D} \cdot d\mathbf{S} = Q$, the electric flux ψ emerging from a point charge is identical with the amount of charge, Figure 4.2.

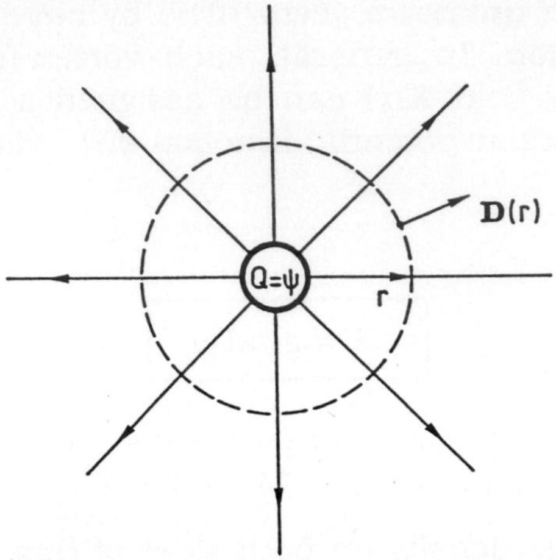

Figure 4.2: Electrostatic field of a positive point-charge (charge with negligible spatial extension).

In a spherical coordinate system the magnitude of the electric flux density at a distance r from a point charge is calculated as the ratio of charge to the area of a sphere with radius r

$$D(r) = \frac{\psi}{S(r)} = \frac{Q}{4\pi r^2} \quad . \tag{4-22}$$

From the flux density the magnitude of the electric field strength at a distance r is found immediately as

$$E(r) = \frac{D(r)}{\varepsilon} = \frac{Q}{4\pi\varepsilon r^2} \quad , \tag{4-23}$$

and upon integrating the field strength at the distance r (allowing $\varphi_o = 0$)

$$\varphi(r) = - \int E(r)\, dr = - \frac{Q}{4\pi\varepsilon} \int \frac{dr}{r^2} = + \frac{Q}{4\pi\varepsilon r} \quad , \tag{4-24}$$

$$\boxed{\varphi(r) = \frac{Q}{4\pi\varepsilon r}} \quad . \tag{4-25}$$

To simplify writing, $1/4\pi r\varepsilon$ is frequently replaced by a *potential coefficient* p_r

$$\boxed{\varphi(r) = p_r\, Q} \qquad (4\text{-}26)$$

Evaluation of the potential becomes more demanding when the charge's position does not coincide with the origin of the coordinate system. Nevertheless, one can visualize again a spherical coordinate system, whose origin coincides with the position of the charge, and calculate the field of the point charge with respect to a radius originating from this new origin. This radius can be represented as the magnitude of the difference of the two position vectors $\mathbf{r}$ and $\mathbf{r}_q$, Figure 4.3.

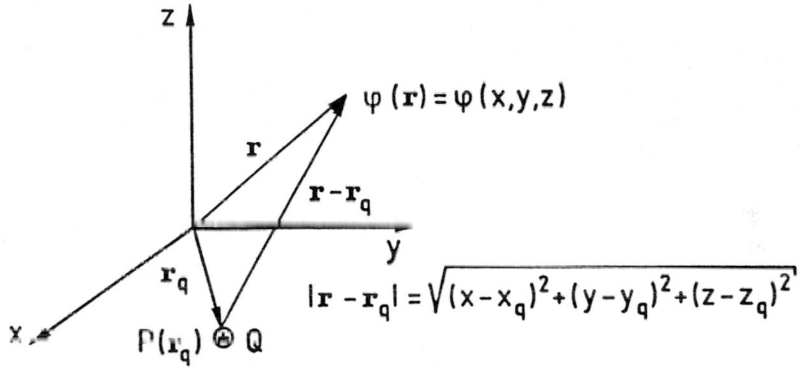

Figure 4.3: Calculating the potential function of a point charge Q at a field point $P(\mathbf{r}_q)$.

Thus, we obtain the potential function of a point charge at $P(\mathbf{r}_q)$

$$\boxed{\varphi(\mathbf{r}) = \frac{Q}{4\pi\varepsilon\,|\mathbf{r} - \mathbf{r}_q|}} \quad \text{or} \quad \boxed{\varphi(\mathbf{r}) = p_{r r_q}\, Q} \qquad (4\text{-}27)$$

Expressing both position vectors $\mathbf{r}_q$ and $\mathbf{r}$ by their components in the respective coordinate system gives the potential function a more familiar analytical form. Vectors are known to be added or subtracted, resp,. by adding or subtracting their components. For instance, in a Cartesian coordinate system one obtains with

$$|\mathbf{r}-\mathbf{r}_q| = \sqrt{(x-x_q)^2 + (y-y_q)^2 + (z-z_q)^2}$$

$$\boxed{\varphi(x,y,z) = \frac{Q}{4\pi\varepsilon\sqrt{(x-x_q)^2 + (y-y_q)^2 + (z-z_q)^2}}} \quad . \qquad (4\text{-}28)$$

The coordinate system is always chosen so as to yield the simplest possible expressions (see A 3).

For several charges at different positions, the potential function of each point charge may be separately calculated irrespective of the existence of other charges. Then, the elementary potential functions are superimposed to give a global potential function integrating all sources in the region considered (*superposition principle*)

$$\varphi(\mathbf{r}) = \frac{Q_1}{4\pi\varepsilon|\mathbf{r}-\mathbf{r}_{q_1}|} + \frac{Q_2}{4\pi\varepsilon|\mathbf{r}-\mathbf{r}_{q_2}|} + \ldots + \frac{Q_n}{4\pi\varepsilon|\mathbf{r}-\mathbf{r}_{q_n}|} \quad , \qquad (4\text{-}29)$$

$$\boxed{\varphi(\mathbf{r}) = \varphi_1(\mathbf{r}) + \varphi_2(\mathbf{r}) + \ldots + \varphi_n(\mathbf{r})} \quad . \qquad (4\text{-}30)$$

4.3.1 Potential Function of a Line Charge

Let ρ_L be a line charge of finite length with uniform charge density,

$$\rho_L = \lim_{\Delta L \to 0} \frac{\Delta Q}{\Delta L} = \frac{dQ}{dL} \quad , \tag{4-31}$$

as depicted in Figure 4.4.

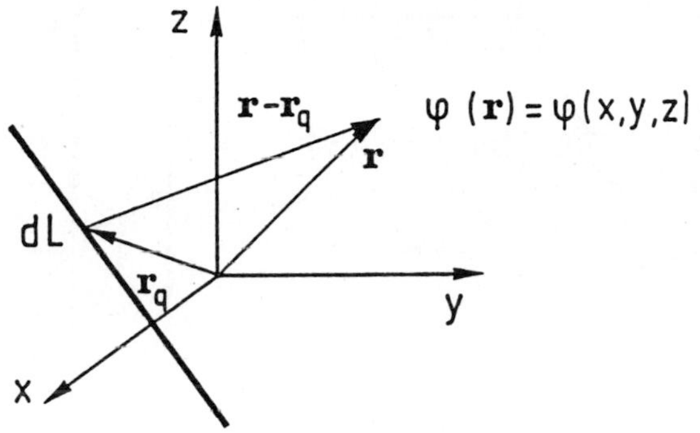

Figure 4.4: Evaluation of the potential function of a line charge with charge density $\rho_L(\mathbf{r}_q)$; the position vector $\mathbf{r}_q$ is the integration variable.

Interpreting an infinitesimally short line element dL as a point charge, its potential function is given in a similar manner as in the previous chapter

$$d\varphi = \frac{dQ}{4\pi\varepsilon|\mathbf{r}-\mathbf{r}_q|} = \frac{\rho_L(\mathbf{r}_q)dL}{4\pi\varepsilon|\mathbf{r}-\mathbf{r}_q|} \quad . \tag{4-32}$$

Summing up the potential functions of all infinitesimal point charges via an integration yields the potential function of a line charge

$$\varphi_L(\mathbf{r}) = \frac{1}{4\pi\varepsilon} \int \frac{\rho_L(\mathbf{r}_q)}{|\mathbf{r}-\mathbf{r}_q|} dL \quad . \tag{4-33}$$

For the solution of this integral one has to specify a particular coordinate system. Consider Cartesian coordinates and let us place the line charge in the z-axis symmetrically about the origin, Figure 4.5 a.

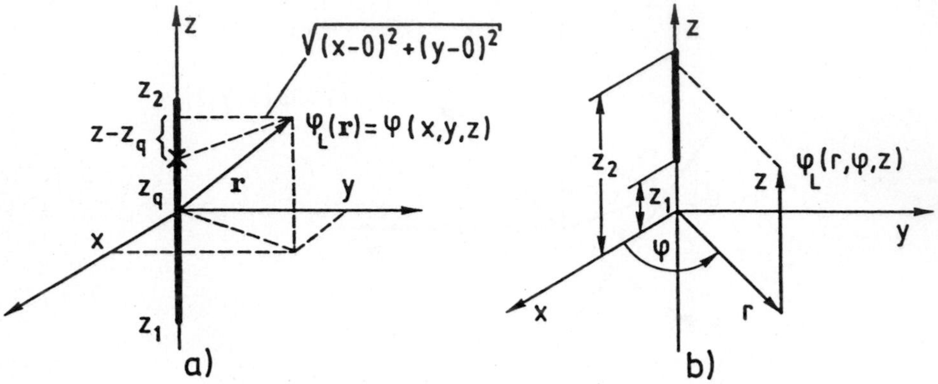

Figure 4.5: Examples of line charges $\rho_L(z)$=const. for Cartesian coordinates (a) and cylindrical coordinates (b).

Denoting the integration variable on the z-axis as z_q (x_q=0, y_q= 0), the potential function is given by

$$\varphi_L(x,y,z) = \frac{1}{4\pi\varepsilon} \int_{z_1}^{z_2} \frac{\rho_L(z_q)}{\sqrt{(x-0)^2 + (y-0)^2 + (z-z_q)^2}}\, dz_q \ , \quad (4\text{-}34)$$

or upon evaluating the integral,

$$\boxed{\varphi_L(x,y,z) = \frac{\rho_L}{4\pi\varepsilon} \ln \frac{z_2 - z + \sqrt{x^2 + y^2 + (z-z_2)^2}}{z_1 - z + \sqrt{x^2 + y^2 + (z-z_1)^2}}} \ . \quad (4\text{-}35)$$

Usually, line charges are specified in cylindrical coordinates. Hence, for the more general arrangement depicted in Figure 4.5b, one obtains

$$\boxed{\varphi_L(r,\varphi,z) = \frac{\rho_L}{4\pi\varepsilon} \ln\left[\frac{z_2 - z + \sqrt{r^2 + (z - z_2)^2}}{z_1 - z + \sqrt{r^2 + (z - z_1)^2}}\right]} \quad . \qquad (4\text{-}36)$$

Similar equations exist for circular line charges referred to as *ring charges.* Mathematically they are more demanding because of the elliptic integrals involved.

4.3.2 Potential Function of a General Charge Configuration

In much the same way as in the previous section, the potential function of a surface charge element with density ρ_S is given by

$$d\varphi = \frac{\rho_S(\mathbf{r}_q)\,dS}{4\pi\varepsilon\,|\mathbf{r}-\mathbf{r}_q|} \quad , \qquad (4\text{-}37)$$

and the potential function of a three-dimensional volume charge element with charge density ρ_V by

$$d\varphi = \frac{\rho_V(\mathbf{r}_q)\,dV}{4\pi\varepsilon\,|\mathbf{r}-\mathbf{r}_q|} \quad . \qquad (4\text{-}38)$$

The most general case of a charge configuration consisting of point charges as well as line, surface, and volume charge densities, is shown in Figure 4.6.

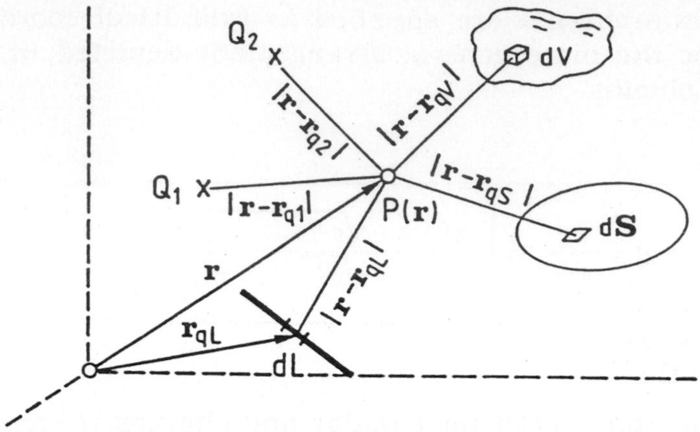

Figure 4.6: Superposition of elementary potentials to a global potential function $\varphi(\mathbf{r})$; integration variable $\mathbf{r}_q$.

By superposition the global potential function is obtained as

$$\varphi(\mathbf{r}) = \frac{1}{4\pi\varepsilon}\left[\sum \frac{Q_v}{|\mathbf{r}-\mathbf{r}_{q_v}|} + \int \frac{\rho_L(\mathbf{r}_q)}{|\mathbf{r}-\mathbf{r}_q|}\,dL + \iint \frac{\rho_S(\mathbf{r}_q)}{|\mathbf{r}-\mathbf{r}_q|}\,dS + \iiint \frac{\rho_V(\mathbf{r}_q)}{|\mathbf{r}-\mathbf{r}_q|}\,dV\right] \tag{4-39}$$

It is to be noted that most technical problems differ from the problem formulation above in that, usually, one does not deal with isolated charges or charge density distributions in an unbounded space, but with charges on electrodes whose surface charge distributions are unknown. These problems are referred to as boundary-value poblems. In this case, one needs to solve the *potential equation* presented in the next section.

We have paid much attention to the potential functions of elementary charge types because the potential equation of technical problems frequently requires a numerical solution, and because some modern computer-aided field calculation methods, for example *charge simulation* and *boundary-element method* ect. are based on the superposition of *elementary potential functions* of various types of image charges (see 9).

4.4 Potential Equations

Apart from the superposition of elementary potential functions of discrete, infinitesimal point charges, a field's potential function can be evaluated by solving the potential equation $\nabla^2\varphi = 0$, subject to the field region's characteristic boundary conditions. In the following sections we will derive the potential equations for electric fields with and without space charges (Laplace and Poisson fields).

4.4.1 Potential Equations for Fields without Space Charges

As was shown in Section 3.2.4, the field region surrounding charges obeys

$$\operatorname{div} \mathbf{D} = 0 \quad . \tag{4-40}$$

Together with the constitutive equation $\mathbf{D} = \varepsilon\mathbf{E}$ and the relation $\mathbf{E} = -\operatorname{grad} \varphi$, we obtain

$$\operatorname{div} \varepsilon\mathbf{E} = 0 \quad \text{or} \quad \operatorname{div} \operatorname{grad} \varphi = 0 \quad . \tag{4-41}$$

For instance, for a scalar field $\varphi(x,y,z)$ in a Cartesian coordinate system the result of the gradient operation yields

$$\operatorname{grad} \varphi(x,y,z) = \frac{\partial\varphi}{\partial x}\mathbf{a}_x + \frac{\partial\varphi}{\partial y}\mathbf{a}_y + \frac{\partial\varphi}{\partial z}\mathbf{a}_z \quad . \tag{4-42}$$

Evaluating this vector field's divergence according to Section 3.2.4, gives

$$\text{div grad } \varphi = \nabla \cdot \nabla\varphi = \left(\frac{\partial}{\partial x}\mathbf{a}_x + \frac{\partial}{\partial y}\mathbf{a}_y + \frac{\partial}{\partial z}\mathbf{a}_z\right)\left(\frac{\partial\varphi}{\partial x}\mathbf{a}_x + \frac{\partial\varphi}{\partial y}\mathbf{a}_y + \frac{\partial\varphi}{\partial z}\mathbf{a}_z\right) \tag{4-43}$$

Considering $\mathbf{a}_\nu \mathbf{a}_\nu = 1$ and discarding the zero terms we obtain

$$\text{div grad } \varphi = \nabla \cdot \nabla\varphi = \nabla^2\varphi = \Delta\varphi = \frac{\partial^2\varphi}{\partial x^2} + \frac{\partial^2\varphi}{\partial y^2} + \frac{\partial^2\varphi}{\partial z^2} = 0\,. \tag{4-44}$$

The operator $\nabla^2 = \Delta$ is called *Laplacian* or *del squared.* Other coordinate systems yield similar equations (see A 3).

The differential equation $\nabla^2\varphi = 0$ is called the *Laplace* equation. Its solutions $\varphi(x,y,z)$ are called *harmonic potentials.* They generate so-called Laplace fields; these are charge-free source field regions.

For a given field the solution of Laplace's equation is obtained taking into account the field's characteristic boundary conditions. One distinguishes between

- *boundary-value problems* of the first type (Dirichlet problem):

  The solution $\varphi(x,y,z)$ must attain certain given potentials on the boundaries (electrode contours).

- *boundary-value problems* of the second type (Neumann problem):

  The first derivative of the solution $\varphi(x,y,z)$ must attain certain given values on the boundary.

In general, an analytic solution of Laplace's equation for given boundary conditions is only manageable for simple problems with a high degree of symmetry, e.g., the electric field of a parallel-plate capacitor without fringe effects, Figure 4.7.

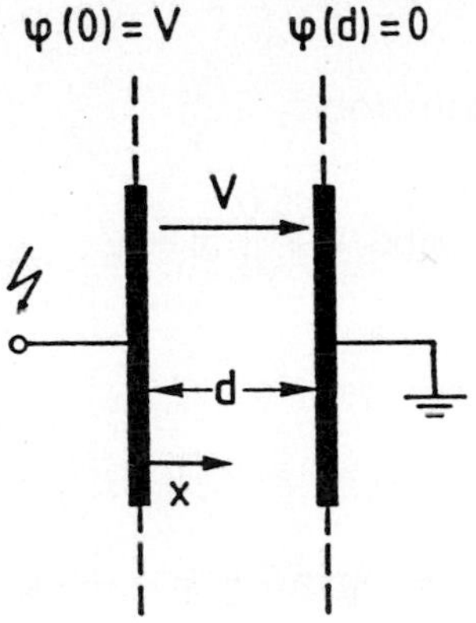

Figure 4.7: Parallel-plate capacitor, fringing fields neglected, e.g. electrode of a guard-ring capacitor.

In the uniform field of a parallel-plate capacitor the field quantities do not vary in y and z ($d\varphi/dz = 0$, $d\varphi/dy = 0$). Therefore, Laplace's equation reduces to

$$\nabla^2\varphi = \frac{\partial^2\varphi}{\partial x^2} + \cancel{\frac{\partial^2\varphi}{\partial y^2}} + \cancel{\frac{\partial^2\varphi}{\partial z^2}} = 0 \quad . \tag{4-45}$$

Integrating $d^2\varphi/dx^2$ once yields

$$\frac{d\varphi}{dx} = K_1 \quad , \tag{4-46}$$

integrating once more

$$\varphi(x) = K_1 x + K_2 \quad . \tag{4-47}$$

The integration constants are determined exploiting the known boundary values

$$\varphi(0) = V = K_1 \cdot 0 + K_2 \rightarrow K_2 = V$$
$$\varphi(d) = 0 = K_1 \cdot d + V \rightarrow K_1 = -\frac{V}{d} \quad . \tag{4-48}$$

Hence, we obtain as solution

$$\varphi(x) = -\frac{V}{d}x + V \quad . \tag{4-49}$$

Taking the gradient,

$$\mathbf{E}(x) = -\text{ grad } \varphi = -\frac{d\varphi}{dx}\mathbf{a}_x \quad , \tag{4-50}$$

we obtain for the x-component of the electric field strength

$$E_x = -\text{ grad } \varphi(x) = -\frac{d\varphi}{dx} = \frac{V}{d} = \text{const.} \quad . \tag{4-51}$$

This result is not exactly surprising and could have been found as well without direct integration of Laplace´s equation. Hence, this example serves only to demonstrate the basic procedure. For a two-dimensional problem the solution effort rapidly increases. Regarding three-dimensional real-world problems without symmetry and various dielectrics, for example the interior of a high-voltage power transformer or three-phase gas-insulated switch-gear, only numerical solution methods employing powerful software and digital computers can cope.

4.4.2 Potential Equations for Electric Fields with Space Charges

As was shown in Section 3.2.3, electric fields with space charges obey

$$\mathrm{div}\,\mathbf{D} = \rho \quad . \tag{4-52}$$

Together with the equation of matter $\mathbf{D} = \varepsilon\,\mathbf{E}$ and with $\mathbf{E}$ = grad φ we obtain

$$\mathrm{div}\,\varepsilon\,\mathbf{E} = \rho \quad \text{or} \quad \mathrm{div\,grad}\,\varphi = -\frac{\rho}{\varepsilon} \quad . \tag{4-53}$$

For instance, for a scalar field $\varphi(x,y,z)$ with space charges, the sequential application of the differential operators grad and div yields

$$\mathrm{div\,grad}\,\varphi = \frac{\partial^2\varphi}{\partial x^2} + \frac{\partial^2\varphi}{\partial y^2} + \frac{\partial^2\varphi}{\partial z^2} = -\frac{\rho}{\varepsilon} \tag{4-54}$$

or

$$\boxed{\nabla^2\varphi = -\frac{\rho}{\varepsilon}} \quad . \tag{4-55}$$

The differential equation $\nabla^2\varphi = -\rho/\varepsilon$ is called *Poisson* equation. Its solutions are composed of two components, a *Newton* - potential function of the space charges ρ, and a *Laplace* - potential function of the arrangement without space charges

$$\varphi(\mathbf{r}) = \varphi_N(\mathbf{r}) + \varphi_L(\mathbf{r}) \quad . \tag{4-56}$$

For example, let us consider the field of a parallel-plate capacitor with uniform space-charge distribution $\varphi(x,y,z)$ = const., as depicted in Figure 4.8.

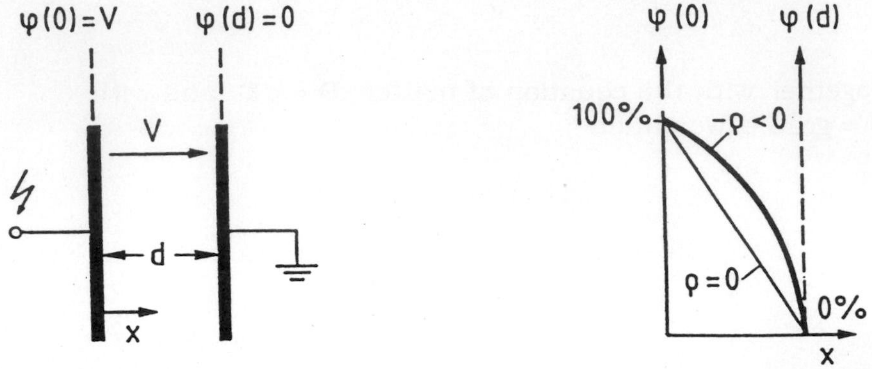

Figure 4.8: Parallel-plate capacitor with uniform space charge distribution ρ(x,y,z) = const.. Fringing fields neglected.

Its field is described by the one-dimensional Poisson equation

$$\frac{d^2\varphi}{dx^2} = -\frac{\rho}{\varepsilon} \quad . \tag{4-57}$$

Two-fold integration yields

$$\varphi(x) = -\frac{\rho}{2\varepsilon}x^2 + K_1 x + K_2 \quad . \tag{4-58}$$

The integration constants are evaluated exploiting the given boundary values

$$\begin{aligned} \varphi(0) &= V = K_2 && \rightarrow \quad K_2 = V \\ \varphi(d) &= 0 = -\frac{\rho}{2\varepsilon}d^2 + K_1 d + V && \rightarrow \quad K_1 = \frac{\rho}{2\varepsilon}d - \frac{V}{d} \quad . \end{aligned} \tag{4-59}$$

Hence, we obtain the solution

$$\varphi(x) = -\frac{\rho}{2\varepsilon}x^2 + \frac{\rho}{2\varepsilon}xd - \frac{V}{d}x + V = -\frac{V}{d}x + V - \frac{\rho}{2\varepsilon}(x^2 - xd)\,, \quad (4\text{-}60)$$

$$\varphi(x) = \underbrace{-\frac{\rho}{2\varepsilon}(x^2 - xd)}_{\varphi_N(x)} \underbrace{-\frac{V}{d}x + V}_{\varphi_L(x)} \quad . \qquad (4\text{-}61)$$

The x- component of the field-strength vector is obtained by a gradient operation

$$E_x = -\,\text{grad}\,\varphi(x) = -\frac{d\varphi(x)}{dx} = \frac{\rho}{\varepsilon}x - \frac{\rho}{2\varepsilon}d + \frac{V}{d}$$

$$E_x = \frac{\rho}{\varepsilon}\left(x - \frac{d}{2}\right) + \frac{V}{d} \quad . \qquad (4\text{-}62)$$

Concerning the difficulties of an analytic solution of Poisson's equation, the aforesaid remarks for the Laplace equation still hold.

To summarize, because curl $\mathbf{X}(\mathbf{r})$ = curl grad $\varphi(\mathbf{r}) = 0$ in vortex-free fields or vortex-free regions of arbitrary vector fields, a vector field $\mathbf{X}(\mathbf{r})$ can be represented as gradient of a scalar potential function $\varphi(\mathbf{r})$. In other words, such fields can be derived by differentiating a scalar potential function $\varphi(\mathbf{r})$ according to the grad rule. Moreover, if the vector field is also source-free, i.e. div $\mathbf{X} = 0$, one obtains $\varphi(\mathbf{r})$ as a solution of Laplace's equation $\nabla^2\varphi = 0$. If the vector field contains sources, i.e. div $\mathbf{X}(\mathbf{r}) = \rho(\mathbf{r})$, one obtains $\varphi(\mathbf{r})$ as a solution of Poisson's equation $\nabla^2\varphi = -\rho/\varepsilon$.

4.5 Electric Vector Potential

When dealing with fields of charge- and current-free regions, for example wave guides, the *electric vector potential* **F** is encountered.

Because of the absence of sources we have div **D**=0 and div **E**=0, and because of the fact that the differential operators div and curl sequentially applied to an arbitrary vector field **F** always yield zero (see A3) we also have div curl **F**=0. Hence, all vortex fields derived via the operator curl from a primary vortex field must satisfy the source-density equation of vortex fields. If we introduce curl **F** into the equation for the source density of **D** (or if we equate div **D**=0 and div curl **F**=0) we formally obtain upon integration, i.e. upon applying the operator $\{\text{div}\}^{-1}$ to both sides of the equation (undoing the div operation)

$$\text{div}\,\mathbf{D} = \text{div}\,\text{curl}\,\mathbf{F} \quad , \tag{4-63}$$

$$\boxed{\mathbf{D} = \text{curl}\,\mathbf{F}} \quad \text{or} \quad \boxed{\text{curl}\,\mathbf{F} = \mathbf{D}} \quad . \tag{4-64}$$

Hence, the electric flux-density vector **D** can be represented as the vortex density of another vector field **F**, referred to as *electric vector potential.*

The symbol $\{\text{div}\}^{-1}$ is a new integral operator inverse to the differential operator div (considering a unique solution which is, in general, provided by problem-specific boundary conditions). This straight-forward procedure is equivalent to the usual statements of the type "...a source-free vector field **D**, characterized by div **D**=0, can, because of div curl **F**=0, be represented by **D** = curl **F** ... ". The interested reader may look up appendix A4 to see what $\{div\}^{-1}$ mathematically stands for.

The electric vector potential has the dimensions As/m (later we shall encounter the magnetic vector potential **A**[Vs/m], see5.3)

Here the question of the uniqueness of **F** must again be raised. As with electrostatic fields, where various scalar potential functions $\varphi(\mathbf{r})$ and $\varphi^*(\mathbf{r})$ can be associated with the same vector field $\mathbf{E}(\mathbf{r})$ (see 4.2),

$$\mathbf{E}(\mathbf{r}) = -\,\mathrm{grad}\varphi(\mathbf{r}) = -\,\mathrm{grad}\,[\varphi(\mathbf{r})+\varphi_0] := -\,\mathrm{grad}\varphi^*(\mathbf{r})\,, \qquad (4\text{-}65)$$

various vector potential functions $\mathbf{F}(\mathbf{r})$ exist for one and the same electric flux density $\mathbf{D}(\mathbf{r})$,

$$\mathbf{D}(\mathbf{r}) = \mathrm{curl}\,\mathbf{F}(\mathbf{r}) = \mathrm{curl}\,[\mathbf{F}(\mathbf{r})+\mathbf{F}_0] := \mathrm{curl}\,\mathbf{F}^*(\mathbf{r}) \quad . \qquad (4\text{-}66)$$

A spatially independent component $\mathbf{F}_0$ = const. vanishes during differentiation and must be supplemented upon integration in form of an integration constant.

In addition to $\mathbf{F}_0$, another term does not appear, and that is the source field component $\mathbf{F}_\Omega$. Its absence is due to the fact that the curl operator applied to the source field $\mathbf{F}_\Omega(\mathbf{r})$ = grad Ω (or to the gradient of any scalar potential function) will always yield zero

$$\mathrm{curl}\,\mathrm{grad}\,\Omega(\mathbf{r}) = \mathrm{curl}\,\mathbf{F}_\Omega(\mathbf{r}) = 0 \quad . \qquad (4\text{-}67)$$

Hence

$$\boxed{\mathbf{D}(\mathbf{r}) = \mathrm{curl}\,\mathbf{F}(\mathbf{r}) = \mathrm{curl}\,[\,\mathbf{F}(\mathbf{r}) + \mathrm{grad}\,\Omega(\mathbf{r})] := \mathrm{curl}\,\mathbf{F}^{**}(\mathbf{r})} \,.$$

(4-68)

In general, the vector-potential function of an electric field may indeed be composed of a vortex field $\mathbf{F}_V(\mathbf{r})$ and a source field $\mathbf{F}_S(\mathbf{r}) = \mathbf{F}_\Omega(\mathbf{r}) = \text{grad}\ \Omega(\mathbf{r})$, a fact already mentioned in section 2.3

$$\mathbf{F}(\mathbf{r}) = \mathbf{F}_V(\mathbf{r}) + \mathbf{F}_S(\mathbf{r}) = \mathbf{F}_V(\mathbf{r}) + \mathbf{F}_\Omega(\mathbf{r}) = \mathbf{F}_V(\mathbf{r}) + \text{grad}\ \Omega(\mathbf{r})\ . \tag{4-69}$$

In other words, the equation curl $\mathbf{F} = \mathbf{D}$ does not define the vector potential function $\mathbf{F}(\mathbf{r})$ uniquely. It merely makes a statement about the vortex field component of the electric vector potential. Uniqueness is only given if there exists also information about the divergence of the source field component $\mathbf{F}_\Omega(\mathbf{r})$ derived from any possible scalar field $\Omega(\mathbf{r})$.

Determination of the integration function, or *gauge function* $\Omega(\mathbf{r})$ (this resembles the determination of the integration constant C of a simple indefinite integral) follows from the problem definition (see also 5.3).

Whereas the electric scalar potential $\varphi(\mathbf{r})$ is valid only for static and quasi-static states of equilibrium, the electric vector potential $\mathbf{F}(\mathbf{r})$ can describe time-varying electric fields.

Instead of starting with div $\mathbf{D} = 0$ one could as well derive a vector potential for div $\mathbf{E} = 0$

$$\boxed{\mathbf{E} = \text{curl}\ \mathbf{F}} \quad \text{or} \quad \boxed{\text{curl}\ \mathbf{F} = \mathbf{E}}\ . \tag{4-70}$$

This vector potential's unit is the volt (V). Regarding its uniqueness, the considerations made above apply in the same way.

Apparently, each general vector field consisting of a vortex and a source component can be assigned an infinite number of scalar and vector potentials (in much the same way as indefinite integrals and differential equations possess an infinite number of solutions). A unique solution is obtained only upon introduction of problem-specific boundary conditions.

Finally, it is to be noted that the potential equations of this chapter qualify only for the calculation of static and quasi-static fields, that is for true or only slowly varying states of equilibrium. For the calculation of rapidly varying fields the more powerful wave equations and so-called retarded potentials must be utilized (see 6).

4.6 Vector Potential of Conduction Fields

In dealing with eddy current problems, the *vector potential of the conduction field* is encountered.

Because of the absence of sources in an electric conduction field – div $\mathbf{J}_c = 0$ (see 3.2.5) and because of the fact that the differential operators div and curl sequentially applied to an arbitrary vector field $\mathbf{T}$ always yield zero (see A 3) – div curl $\mathbf{T}=0$ – the flux density $\mathbf{J}_c$ of the electric conduction field (current density) may be represented as the vortex density of a vortex field $\mathbf{T}$, the *vector potential of the conduction field.* Equating both equations

$$\text{div}\,\mathbf{J}_c = \text{div curl}\,\mathbf{T} \quad , \tag{4-71}$$

and integrating both sides (undoing the div operation, $\{\text{div}\}^{-1}$) yields

$$\boxed{\mathbf{J}_c = \text{curl}\,\mathbf{T}} \quad \text{or} \quad \boxed{\text{curl}\,\mathbf{T} = \mathbf{J}_c} \quad . \tag{4-72}$$

Here the question of uniqueness of **T** must again be raised. As with electrostatic fields where various scalar potential functions $\varphi(\mathbf{r})$ and $\varphi^*(\mathbf{r})$ can be associated with the same vector field $\mathbf{E}(\mathbf{r})$ (see 4.2),

$$\mathbf{E}(\mathbf{r}) = -\,\mathrm{grad}\,\varphi(\mathbf{r}) = -\,\mathrm{grad}\,[\varphi(\mathbf{r}) + \varphi_0] := -\,\mathrm{grad}\,\varphi^*(\mathbf{r}) \quad , \qquad (4\text{-}73)$$

various vector potential functions $\mathbf{T}(\mathbf{r})$ exist for one and the same current-density field $\mathbf{J}_c$,

$$\mathbf{J}_c(\mathbf{r}) = \mathrm{curl}\,\mathbf{T}(\mathbf{r}) = \mathrm{curl}\,[\mathbf{T}(\mathbf{r}) + \mathbf{T}_0] := \mathrm{curl}\,\mathbf{T}^*(\mathbf{r}) \quad . \qquad (4\text{-}74)$$

A spatially independent component $\mathbf{T}_0$ = const. vanishes during differentiation and must be supplemented upon integration in form of an integration constant.

In addition to $\mathbf{T}_0$ another term does not appear, and that is the source field component $\mathbf{T}_\Omega(\mathbf{r})$. Its absence is due to the fact that the curl operator applied to the source field $\mathbf{T}_\Omega(\mathbf{r}) = \mathrm{grad}\,\Omega$ (or the gradient of any scalar potential function) will always yield zero

$$\mathrm{curl}\,\mathrm{grad}\,\Omega(\mathbf{r}) = \mathrm{curl}\,\mathbf{T}_\Omega(\mathbf{r}) = 0 \quad . \qquad (4\text{-}75)$$

Hence,

$$\boxed{\mathbf{J}_c(\mathbf{r}) = \mathrm{curl}\,\mathbf{T}(\mathbf{r}) = \mathrm{curl}\,[\,\mathbf{T}(\mathbf{r}) + \mathrm{grad}\,\Omega(\mathbf{r})] := \mathrm{curl}\,\mathbf{T}^{**}(\mathbf{r})} \quad . \qquad (4\text{-}76)$$

In general, the vector-potential function of a conduction field may indeed be composed of a vortex field $\mathbf{T}_V(\mathbf{r})$ and a source field $\mathbf{T}_S(\mathbf{r}) = \mathbf{T}_\Omega(\mathbf{r}) = \text{grad}\ \Omega(\mathbf{r})$, (s. also 4.5). In other words, the equation curl $\mathbf{T} = \mathbf{J}_c$ does not define the vector-potential function $\mathbf{T}(\mathbf{r})$ uniquely. It merely makes a statement about the vortex field component of the electric vector potential. Uniqueness is only given if there exists also information about the divergence of the source field component $\mathbf{T}_\Omega(\mathbf{r})$ derived from any possible scalar field $\Omega(\mathbf{r})$. Determination of the integration function, or *gauge function* $\Omega(\mathbf{r})$ (this resembles the determination of the integration constant C of a simple indefinite integral) follows from the problem definition (see also 5.3).

Whereas the electric scalar potential $\varphi(\mathbf{r})$ is valid only for static and quasi-static states of equilibrium, the vector potential $\mathbf{T}(\mathbf{r})$ of the conduction field can describe time-varying electric fields.

5 Potential and Potential Function of Magnetostatic Fields

Magnetostatic fields exist inside and outside of conductors carrying direct current, in the surroundings of permanent magnets, and around electric fields changing at a constant rate (magnetostatic fields of a constant displacement-current density $\dot{\mathbf{D}}$). Here, we will illustrate the potential concept for magnetostatic fields resulting from dc-currents.

5.1 Magnetic Scalar Potential

A magnetic scalar potential $\varphi_m(\mathbf{r})$ can be derived for vortex-free regions of magnetostatic fields from a given current distribution (see Chapter 4), like an electrostatic field, where a scalar potential function $\varphi(\mathbf{r})$ can be derived from a given charge distribution by virtue of the absence of vortices,

$$\boxed{\varphi_m = - \int \mathbf{H} \cdot d\mathbf{r} + \varphi_{m_0}} \quad . \qquad (5\text{-}1)$$

The magnetic scalar potential's unit is ampere (A).

In formal analogy with the electric field, the magnetic field strength is calculated as the gradient of this scalar potential (the negative sign is arbitrary and serves only to emphasize the analogy, see also 4.1)

$$\boxed{\mathbf{H} = -\operatorname{grad} \varphi_m} \quad . \tag{5-2}$$

Taking the vortex density of both sides gives

$$\operatorname{curl} \mathbf{H} = \operatorname{curl} \operatorname{grad} \varphi_m = 0 \,, \tag{5-3}$$

because the differential operators curl and grad sequentially applied to an arbitrary scalar field always yield zero (see 4.2). Moreover, this implies that only such functions as $\mathbf{H}(x,y,z)$ can be represented as the gradient of a scalar field which obey curl $\mathbf{H} = 0$. The latter constraint is complied with only in current-free regions of magnetostatic vortex fields, i.e. outside current carrying conductors. According to Section 3.2.2, curl $\mathbf{H} = \mathbf{J}$ applies to the interior of the conductors (see also Chapter 6). Hence, use of the magnetic scalar potential is restricted to *vortex-free*, or *current-free*, regions.

As in the electrostatic field, a reference potential φ_{m_0} can be arbitrarily chosen. However, this potential need not guarantee uniqueness. Evaluating the magnetic potential-difference between two points $P(\mathbf{r}_1)$ and $P(\mathbf{r}_2)$ from the familiar line integral frequently used in electrostatic fields (see 4.2),

$$\boxed{\varphi_{m_{12}} = -\int_{\mathbf{r}_2}^{\mathbf{r}_1} \mathbf{H} \cdot d\mathbf{r}} \quad , \tag{5-4}$$

the integration path must not enclose a current because the integral

$$\oint \mathbf{H} \cdot d\mathbf{r} \qquad (5\text{-}5)$$

is not path-independent. Whereas in electrostatic fields we always have

$$\oint \mathbf{E} \cdot d\mathbf{r} = 0 \ , \qquad (5\text{-}6)$$

in magnetostatic fields we will find

$$\oint \mathbf{H} \cdot d\mathbf{r} = NI \ , \qquad (5\text{-}7)$$

this result being dependent on the number N of encirclements. Choosing different paths between two points may yield different values for the potential difference. Although we will not elaborate on the explicit calculation of the scalar potential, the reader will get an idea of the procedure when we derive the magnetic vector potential in section 5.3.

5.2 Potential Equation of Magnetic Scalar Potentials

In this section we will show that a similar potential equation $\nabla^2\varphi_m = 0$ exists for the magnetic scalar potential as exists for the scalar potential of an electrostatic field.

According to Section 3.2.4, magnetic fields always obey

$$\operatorname{div} \mathbf{B} = 0 \ . \qquad (5\text{-}8)$$

Together with $\mathbf{H} = -\operatorname{grad}\varphi_m$ and the constitutive relation $\mathbf{B} = \mu\mathbf{H}$, we obtain

$$\operatorname{div}\mu\mathbf{H} = 0 \quad \text{or} \quad \operatorname{div}\operatorname{grad}\varphi_m = 0 \ . \tag{5-9}$$

For instance, for a scalar field $\varphi_m(x,y,z)$ in a Cartesian coordinate system, the result of the gradient operation will be

$$\boxed{\operatorname{grad}\varphi_m(x,y,z) = \frac{\partial\varphi_m}{\partial x}\mathbf{a}_x + \frac{\partial\varphi_m}{\partial y}\mathbf{a}_y + \frac{\partial\varphi_m}{\partial z}\mathbf{a}_z} \ . \tag{5-10}$$

Taking the divergence of this vector field according to equation (4-43) of Section 4.4.1, we finally obtain a potential equation $\nabla^2\varphi_m = 0$, having the same form as $\nabla^2\varphi = 0$.

$$\operatorname{div}\operatorname{grad}\varphi_m = \frac{\partial^2\varphi_m}{\partial x^2} + \frac{\partial^2\varphi_m}{\partial y^2} + \frac{\partial^2\varphi_m}{\partial z^2} = 0 \ , \tag{5-11}$$

or, more compactly,

$$\boxed{\nabla^2\varphi_m = \Delta\varphi_m = 0} \ . \tag{5-12}$$

Hence, calculation of magnetic fields outside current-carrying conductors can be accomplished with the same methods employed in electrostatic source fields. Upon integration of $\nabla^2\varphi_m = 0$, i.e. solving the potential equation for given boundary conditions, one obtains the magnetic field strength $\mathbf{H}(\mathbf{r})$ by differentiating the scalar magnetic potential $\varphi_m(\mathbf{r})$. A typical problem that can be solved with the scalar potential is the evaluation of the magnetostatic shielding efficiency of a ferromagnetic shield.

5.3 Magnetic Vector Potential

As has been mentioned above, the magnetic scalar potential fails inside current-carrying conductors, for example in calculating shielding efficiencies for *time-varying* magnetic vortex fields or in treating wave problems in which the magnetic effects of the displacement currents outside current-carrying conductors cannot be neglected (see 6.1). Fortunately, in addition to the magnetic scalar potential $\varphi_m(\mathbf{r})$, a *magnetic vector potential* $A(\mathbf{r})$ can be defined which is also applicable in regions with current flow.

Because of the absence of sources in magnetic vortex fields we have div **B** = 0 and, since the differential operators div and curl sequentially applied to an arbitrary vector field **A** always yield zero, we have div curl **A** = 0 (see A3). Upon equating and integrating both equations (undoing the div operation, $\{\mathrm{div}\}^{-1}$) the flux density **B** can be formally represented as the vortex density of a vortex field **A** (see also 4.2, 4.5 and A4). Thus,

$$\mathrm{div}\,\mathbf{B} = \mathrm{div}\,\mathrm{curl}\,\mathbf{A} \quad , \tag{5-13}$$

followed by $\{\mathrm{div}\}^{-1}$, yields

$$\boxed{\mathbf{B} = \mathrm{curl}\,\mathbf{A}} \quad \text{or} \quad \boxed{\mathrm{curl}\,\mathbf{A} = \mathbf{B}} \quad . \tag{5-14}$$

The differential operation div curl **A** = 0 can be understood not only formally but also conceptually. The vortex densities curl **A** of a magnetic vector potential **A** are the flux-density lines **B** of the magnetic field considered. These are always solenoidal (see Section 3.2.4) Figure 5.1.

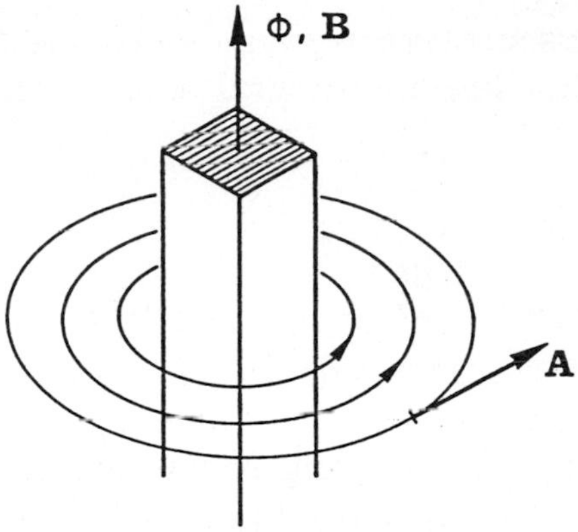

Figure 5.1: Illustration of the vector potential **A** as a vortex field with vortex density curl **A** = **B**. **A** and **B** obey the right-hand rule.

In Section 2.2 we have already seen that vortices and vortex densities, are, respectively, solenoidal tubes and lines whose divergences are clearly zero. With this in mind, the divergence of the vortex density of the vortex field **A** must also be zero, i.e. div curl **A** = div **B** = 0. The magnetic vector potential **A** is a vortex field, its vortices are flux tubes; its vortex densities are flux-density lines **B**. The *vortex strength* of the vector potential is calculated from

$$\boxed{\oint_C \mathbf{A} \cdot d\mathbf{r} = \phi} \quad . \qquad (5\text{-}15)$$

and gives the magnetic flux.

Its *vortex density* is by definition

$$\boxed{\text{curl}\,\mathbf{A} = \mathbf{B}} \quad . \qquad (5\text{-}16)$$

In order to underscore the linkage between these equations, recall the induction law in integral and differential form (see 3.1.1, 3.3.1)

$$\oint \mathbf{E} \cdot d\mathbf{r} = -\frac{d\phi}{dt} , \qquad \text{curl } \mathbf{E} = -\frac{\partial \mathbf{B}}{\partial t} \qquad (5\text{-}17)$$

which result from differentiating the preceding equations with respect to time and recognizing $d\mathbf{A}/dt = -\mathbf{E}_V$ (see 6.2.1).

For the magnetic vector potential, the question of uniqueness must again be raised. Recall that in the electrostatic field different scalar potential functions $\varphi(\mathbf{r})$ may be associated with the same vector field $\mathbf{E}(\mathbf{r})$

$$\mathbf{E}(\mathbf{r}) = -\text{ grad } \varphi(\mathbf{r}) = -\text{ grad } \left(\varphi(\mathbf{r}) + \varphi_0\right) := -\text{ grad } \varphi^*(\mathbf{r}) . \qquad (5\text{-}18)$$

In like manner, in magnetostatic fields various vector potential functions $\mathbf{A}(\mathbf{r})$ may exist for one and the same magnetic flux-density field $\mathbf{B}(\mathbf{r})$

$$\mathbf{B}(\mathbf{r}) = \text{curl } \mathbf{A}(\mathbf{r}) = \text{curl } \left(\mathbf{A}(\mathbf{r}) + \mathbf{A}_0\right) := \text{curl } \mathbf{A}^*(\mathbf{r}) , \qquad (5\text{-}19)$$

since here too, a spatially independent component $\mathbf{A}_0$ vanishes during the curl operation.

In addition to $\mathbf{A}_0$, another term does not appear, and that is the source field component $\mathbf{A}_\Omega(\mathbf{r})$. Its absence is due to the fact that the curl operator applied to the source field $\mathbf{A}_\Omega(\mathbf{r}) = \text{grad } \Omega(\mathbf{r})$ (or to the gradient of any scalar potential function) will always yield zero,

$$\text{curl grad } \Omega\,(\mathbf{r}) = \text{curl } \mathbf{A}_\Omega(\mathbf{r}) = 0 \quad . \qquad (5\text{-}20)$$

Hence,

$$\mathbf{B}(\mathbf{r}) = \text{curl } \mathbf{A}(\mathbf{r}) = \text{curl } (\mathbf{A}(\mathbf{r}) + \text{grad } \Omega(\mathbf{r})) := \text{curl } \mathbf{A}^{**}(\mathbf{r}) \qquad (5\text{-}21)$$

In general, the vector-potential function of a magnetic field may indeed be composed of a vortex field $\mathbf{A}_V(\mathbf{r})$ and a source field $\mathbf{A}_S(\mathbf{r}) = \mathbf{A}_\Omega(\mathbf{r}) = \text{grad } \Omega(\mathbf{r})$. Assuming that the magnetic vector potential $\mathbf{A}(\mathbf{r})$ of a magnetic field is a pure vortex field as in Figure 5.1, we can specify div $\mathbf{A} = 0$. In other words, a vector field $\mathbf{A}_\Omega(\mathbf{r})$ with source character does not exist, and grad $\Omega(\mathbf{r}) = 0$ (*Coulomb* gauge). When we will be discussing time-varying magnetic fields we will reconsider this choice (see Chapter 6).

Finally, we will discuss the evaluation of the magnetic vector potential from a given current distribution. As in the electrostatic field where the potential function of a line charge was calculated by first evaluating the contribution of an infinitesimal line element dL or its point charge dQ (see 4.3), we start with the contribution of an infinitesimal current filament dI to find the magnetic field of a current-carrying conductor loop, Figure 5.2.

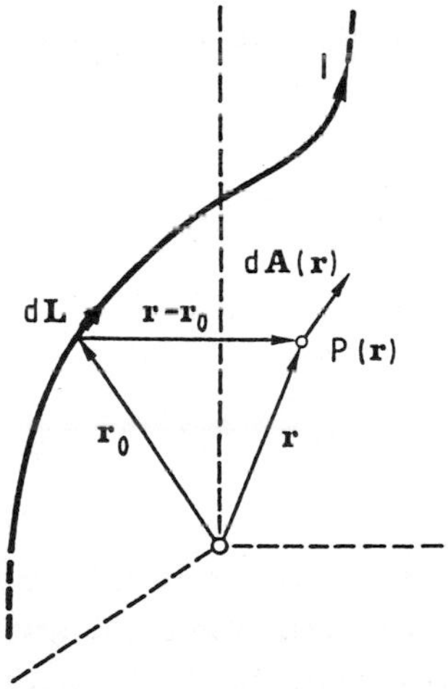

Figure 5.2: Calculation of the magnetic vector potential of a current loop. Integration variable $\mathbf{r}_0$.

The vector potential of the conductor element $d\mathbf{L}$ at the field point P is calculated from

$$d\mathbf{A}(\mathbf{r}) = \frac{\mu_0 I\, d\mathbf{L}}{4\pi|\mathbf{r}-\mathbf{r}_0|} \quad . \tag{5-22}$$

Here, one is tempted to establish an analogy between the products $I d\mathbf{L}$ and $\rho_L dL$. However, the comparison is not perfect. While an infinitesimal charge dQ may exist, an independent infinitesimal current element will not, because currents exist only in closed circuits (div $\mathbf{J}$ = 0). Hence, the equation above makes sense only upon integration about a closed integration path

$$\boxed{\mathbf{A}_L(\mathbf{r}) = \oint \frac{\mu_0 I\, d\mathbf{L}}{4\pi|\mathbf{r}-\mathbf{r}_0|}} \quad . \tag{5-23}$$

Now the analogy with the scalar potential of a line charge becomes more reasonable

$$\boxed{\varphi(\mathbf{r}) = \int \frac{\rho_L\, dL}{4\pi\varepsilon|\mathbf{r}-\mathbf{r}_0|}} \quad . \tag{5-24}$$

For currents distributed over a surface with surface-current density $\mathbf{K}(\mathbf{r}_0)$ one defines infinitesimal surface-current products $\mathbf{K}(\mathbf{r}_0)dS$, and for spatial current distributions with current density $\mathbf{J}(\mathbf{r}_0)$ we have infinitesimal volume-current products $\mathbf{J}(\mathbf{r}_0)dV$, where the element length dL is implicitly included in dS and dV,

$$\boxed{\mathbf{A}_S(\mathbf{r}) = \int_S \frac{\mu_0 \mathbf{K}(\mathbf{r}_0)\,dS}{4\pi|\mathbf{r}-\mathbf{r}_0|}} \quad , \tag{5-25}$$

$$\boxed{\mathbf{A}_V(\mathbf{r}) = \int_V \frac{\mu_0 \mathbf{J}(\mathbf{r}_0)\,dV}{4\pi|\mathbf{r}-\mathbf{r}_0|}} \quad . \tag{5-26}$$

It should be noted that in the line integral, the vector quality of the integrand is expressed by the line element **dL** because the current is by nature a flux, i.e. a scalar.

In a Cartesian coordinate system the three component vectors of the magnetic vector potential are given by

$$A_x = \int_V \frac{\mu_0 J_x(\mathbf{r}_0)}{4\pi|\mathbf{r}-\mathbf{r}_0|}\,dV$$

$$A_y = \int_V \frac{\mu_0 J_y(\mathbf{r}_0)}{4\pi|\mathbf{r}-\mathbf{r}_0|}\,dV$$

$$A_z = \int_V \frac{\mu_0 J_z(\mathbf{r}_0)}{4\pi|\mathbf{r}-\mathbf{r}_0|}\,dV \quad . \tag{5-27}$$

Hence, three integrals need to be evaluated.

5.4 Potential Equation of the Magnetic Vector Potential

In addition to calculating the vector potential from a given current or current-density distribution, the vector-potential function $\mathbf{A}(\mathbf{r})$ can also be understood as a solution to a vector potential-equation.

The vortex density of a magnetic vortex-field is given by (see 3.2.2)

$$\text{curl}\,\mathbf{H} = \mathbf{J} \quad . \tag{5-28}$$

Substituting for **H** the magnetic vector potential via

$$\mathbf{B} = \mu\mathbf{H} = \text{curl}\,\mathbf{A} \quad \rightarrow \quad \mathbf{H} = \frac{1}{\mu}\,\text{curl}\,\mathbf{A}\,, \tag{5-29}$$

yields

$$\text{curl curl}\,\mathbf{A} = \mu\mathbf{J} \quad . \tag{5-30}$$

In order to avoid obscuring the derivation of the vector potential-equation, the reader must accept without evidence that the repeated application of the differential operator curl to an arbitrary vector field is equivalent to the difference of a cascaded div and grad operation and the Laplacian operated on the same vector (see A3). Thus

$$\text{curl curl}\,\mathbf{X} = \text{grad div}\,\mathbf{X} - \nabla^2\mathbf{X} \quad . \tag{5-31}$$

With this in mind, the above equation can be modified to

$$\text{grad div } \mathbf{A} - \nabla^2\mathbf{A} = \mu\mathbf{J} \quad . \tag{5-32}$$

With the previously made assumption that the vortices of a vortex field are also solenoidal and that the sources which could provide an additional contribution to the magnetic vector-potential should not exist, we will have div **A** = 0. Hence, the equation above reduces to the Poisson vector potential-equation of the magnetic vector potential

$$\boxed{\nabla^2\mathbf{A} = -\mu\mathbf{J}} \quad . \tag{5-33}$$

The reader may recognize the formal analogy with the symbolic notation of the scalar potential-equation of an electric field with sources (Poisson equation, see 4.4.2)

$$\boxed{\nabla^2\varphi = -\frac{\rho}{\varepsilon}} \quad . \tag{5-34}$$

It is to be noted that, in general, application of the Laplacian to a vector field again yields a vector field.

By virtue of the superposition principle, the vector differential equation $\nabla^2\mathbf{A} = -\mu\mathbf{J}$ can be split into three simple scalar potential-equations, one for each of the three coordinates

$$\nabla^2 A_x = -\mu J_x$$
$$\nabla^2 A_y = -\mu J_y$$
$$\nabla^2 A_z = -\mu J_z \quad . \qquad (5\text{-}35)$$

Each of these equations formally represents a scalar Poisson equation.

Finally, it should be pointed out that the evaluation of the magnetic vector potential from its potential equation is only possible for magnetostatic or quasi-static fields and under the assumption div $\mathbf{A} = 0$ (see 5.3)

6 Classification of Electric and Magnetic Fields

One can distinguish between time-constant (static) and time-varying fields. Moreover, the latter can be subdivided into slowly and rapidly varying fields, Figure 6.1

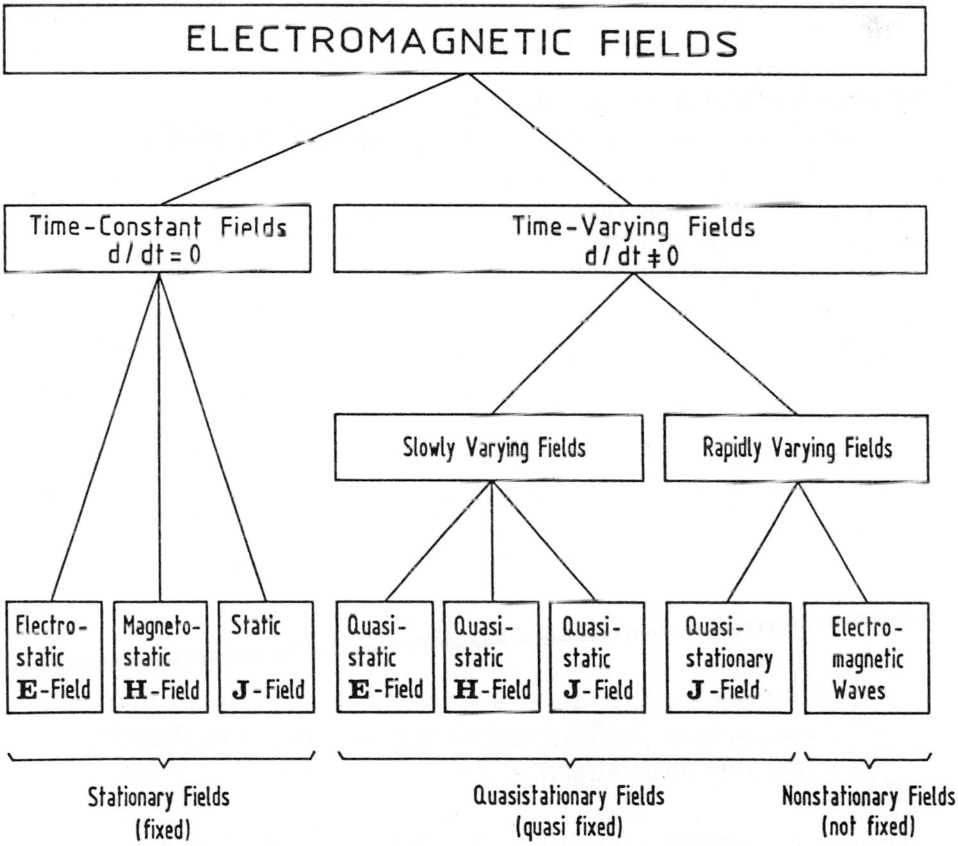

Figure 6.1: Classification of electric and magnetic fields.

Static fields are characterized by fixed, time-constant charges or by charges in motion with constant charge density and speed. Hence, this category also includes all fields resulting from dc voltages or dc currents. The field quantities **E**, **D**, **H**, **B**, **J** and ρ are not functions of time, that is $d/dt = 0$. All time-varying terms in Maxwell's equations vanish. No mutual coupling exists between electric and magnetic fields. Both types of fields may exist individually or simultaneously, without influencing each other.

Static fields can be subdivided into *electrostatic fields*, *magnetostatic fields*, and *static conduction fields*. These fields are generically called *stationary fields*, these are locally fixed fields whose field quantities are bound to the position or region of their generation. Hence, these fields are only functions of space variables.

In time-varying fields all field quantities can be time dependent. Moreover, due to Faraday's law, time-varying magnetic fields produce in their environment electric vortex-fields, and, due to Ampere's law, time-varying electric fields or their displacement currents produce magnetic vortex-fields. Time-varying magnetic and electric fields are always mutually coupled, therefore they are often referred to as electromagnetic fields. To what degree coupling manifests itself will be discussed in detail shortly.

Time-varying fields are divided into slowly and rapidly varying fields. Regarding slowly varying fields, at any instant a snap shot view

of $\mathbf{E}(\mathbf{r},t_v)$ coincides with the *electrostatic field* of a respective static charge distribution,
of $\mathbf{H}(\mathbf{r},t_v)$ coincides with the *magnetostatic field* of a respective dc current, and
of $\mathbf{J}(\mathbf{r},t_v)$ coincides with the *static conduction field* of a respective dc-current (time-varying conduction field without skin effect).

Therefore, space and time dependence are obviously decoupled, allowing slowly varying fields to be treated with the same methods as applied to static or stationary fields. Fields with decoupled space and time dependence are called *quasi-static*.

Concerning *quasi-static electric fields*, the reason for their coincidence with electrostatic fields is understood from the fact that in slowly varying electric fields the time derivative of the electric flux density **D**, that is the displacement current density

$$\mathbf{J}_d = \dot{\mathbf{D}} \quad , \tag{6-1}$$

attains only minor magnitudes, and that the time rate-of-change of the displacement current's magnetic field is so small that the electric vortex-field induced by it does not manifest itself noticeably (see 6.2.1).

Concerning *quasi-static magnetic fields* and *time-varying conduction fields without skin effect*, the coincidence with their respective static fields is explained by the fact that for practically all reasonable frequencies (see 3.1.2) the displacement current in a conductor and its magnetic field can always be neglected when compared to the conduction current and its magnetic field. Moreover, variations are supposed to be so slow that the electric vortex field in the conductor (induced by the conduction current's magnetic field) does not interfere significantly with the current-driving primary field (applied field), i.e.

$$\oint \mathbf{E} \cdot d\mathbf{r} \approx 0 \tag{6-2}$$

or

$$\text{curl}\, \mathbf{E} \approx 0 \tag{6-3}$$

inside the conductor (see 6.2.4). Regarding the entire circuit formed by the conductor,

$$\oint \mathbf{E} \cdot d\mathbf{r} \tag{6-4}$$

need not vanish necessarily (self induced voltage - see Section 5.3). Of course, the external magnetic field of the parasitic external electric field's displacement current is also neglected compared to the magnetic field of the conductor current.

In rapidly varying fields, the displacement-current density $\dot{\mathbf{D}}$ increases in proportion to the time rate-of-change or frequency of the electric field. This results in different effects for conductors and dielectrics. In conductors, the displacement current is as negligible as ever. However, the electric vortex-field induced in the conductor by the magnetic field of the conduction-current component becomes substantial. Super position of the induced electric vortex-field and the applied electric field driving the conduction current results in the skin effect observed at high frequencies (see 6.2.4). The conduction field with skin effec exhibits a different field distribution (flow profile) than the static field and the time-varying conduction field without skin effect. Now the field quantities in the conductor are functions of space and time. Their mathematical modeling is accomplished employing the *diffusion equation* (see 6.2.4).

The following fields

- quasi-static electric field
- quasi-static magnetic field
- time-varying conduction field without skin effect
- time-varying conduction field with skin effect

have in common the fact that the *inducing effect of the displacement current's magnetic field* is neglected, and in conductors the displacement current itself is neglected. Since these fields, despite their space and time dependence, do not possess wave character and because they are also locally fixed as static fields are, they are generically called *quasi-stationary*. Because space and time dependence of static fields are decoupled, these fields lack any propagation quality. Regarding the conduction field with skin effect, "propagation" as described by the diffusion equation is restricted to the interior of the locally fixed conductors (see 6.2.4).

Quasi-static fields represent a subset of quasi-stationary fields. Characterizing quasi-stationary fields by the plain requirement for a negligible displacement current frequently causes confusion because the displacement current is the dominant current component of capacitors in ac circuits. Nevertheless, the fields in these capacitors are considered quasi-static or quasi-stationary. Hence it is not the displacement current that is neglected in dielectrics but merely the electric vortex-field resulting from the displacement current´s magnetic field; in other words, the displacement current´s induction effect is neglected (see 6.2.1).

Finally, there remain rapidly varying fields, for which the displacement current´s induction effect in dielectrics must definitely be considered. In examining such fields outside a conductor, the field quantities **E**, **D**, **H**, **B** are to be taken as functions of space *and* time as encountered in the conduction field with skin effect. Electric and magnetic fields are strictly coupled and expand as electromagnetic waves into space. Rapidly varying fields with wave quality are called *non-stationary fields*; they are not bound to a particular region except at their origin. Mathematically, they are described by *wave equations* (see 6.3).

The following sections elaborate on the various fields regarding their time and space characteristics. In addition, formal criteria will be represented which will allow quantitive distinction among quasistationary fields, non-stationary fields, and electromagnetic waves.

6.1 Stationary Fields

6.1.1 Electrostatic Fields

Drawing on earlier discussions, electrostatic fields exist in the nonconducting environment of electric charges at rest, for example in the vicinity of electrostatically charged dielectrics or around electrodes connected to a dc-voltage source. Because $\sigma = 0$, currents do not exist, hence, compared with general stationary fields, we have in addition $\mathbf{J} = 0$. Electrostatic fields are *vortex-free* or *irrotational* or *conservative*, curl $\mathbf{E}(\mathbf{r}) = 0$. Their sources are electric charges with a source density given by

$$\text{div}\,\mathbf{D}(\mathbf{r}) = \rho(\mathbf{r}) \quad . \tag{6-5}$$

Electrostatic fields belong to the group of source fields (see 2.1). Regions between charges or electrodes are *vortex-* and *source-free*. Electrostatic fields are governed by the following system of equations, Table 6.1.

ELECTROSTATIC FIELD EQUATIONS	
INTEGRAL FORM	DIFFERENTIAL FORM
$\oint_C \mathbf{E} \cdot d\mathbf{r} = 0$	$\text{curl}\,\mathbf{E} = 0$
$\oint_S \mathbf{D} \cdot d\mathbf{S} = Q$	$\text{div}\,\mathbf{D} = \rho$
$\mathbf{D} = \varepsilon \mathbf{E}$	$\nabla^2 \varphi = -\dfrac{\rho}{\varepsilon}$

Table 6.1: Electrostatic field equations.

In charge-free, (source-free regions), Q=0 and $\rho = 0$ in the equations above.

Calculation of electrostatic fields can be accomplished either directly from Gauss´s law or indirectly from the potential function $\varphi(\mathbf{r})$ and subsequent differentiation (gradient operation). The potential function $\varphi(\mathbf{r})$ is evaluated either directly from a given charge distribution or as solution of the potential equations for electrostatic fields with or without space charges.

6.1.2 Magnetostatic Fields

As discussed earlier, magnetostatic fields exist inside and outside direct-current carrying conductors and in the environment of permanent magnets. Moreover, a constant displacement current density

$$\dot{\mathbf{D}} = \varepsilon \dot{\mathbf{E}} \tag{6-6}$$

of a time-varying electric field with constant time rate-of-change produces a magnetostatic field, which, however, shall not be considered here. Magnetostatic fields are *source-free*, since magnetic charges do not exist, so

$$\operatorname{div} \mathbf{B}(\mathbf{r}) = 0 \ . \tag{6-7}$$

Magnetostatic fields are caused by moving charges or the equivalent dc currents (vortices) with a vortex density

$$\operatorname{curl} \mathbf{H}(\mathbf{r}) = \mathbf{J}_C(\mathbf{r}) \tag{6-8}$$

(see 2.2). In *current-free* regions, i.e. outside or between conductors, magnetostatic fields are *vortex-* and *source-free*. Magnetostatic fields are governed by the following system of equations, Table 6.2.

MAGNETOSTATIC FIELD EQUATIONS		
INTEGRAL FORM	DIFFERENTIAL FORM	
$\oint_C \mathbf{H} \cdot d\mathbf{r} = I$	curl $\mathbf{H} = \mathbf{J}$	
$\oint_S \mathbf{B} \cdot d\mathbf{S} = 0$	div $\mathbf{B} = 0$	
$\mathbf{B} = \mu \mathbf{H}$	$\nabla^2 \varphi_m = 0$	$\nabla^2 \mathbf{A} = -\mu \mathbf{J}$

Table 6.2: Magnetostatic field equations.

Calculation of magnetostatic fields is accomplished either directly from Ampere´s law or indirectly via a scalar potential function $\varphi_m(\mathbf{r})$ or a vector potential function $\mathbf{A}(\mathbf{r})$ and subsequent differentiation

$$\mathbf{H} = -\operatorname{grad} \varphi_m \quad \text{and} \quad \mathbf{B} = \mu \mathbf{H} = \operatorname{curl} \mathbf{A} . \tag{6-9}$$

The scalar potential function $\varphi_m(\mathbf{r})$ is evaluated either from a current distribution or from the scalar potential equation $\nabla^2 \varphi_m = 0$ (applicable only outside of current-carrying conductors). The vector potential function $\mathbf{A}(\mathbf{r})$ (*magnetic vector potential*) is evaluated either from a given current distribution or from the vector potential equation of the magnetic field

$$\nabla^2 \mathbf{A} = -\mu \mathbf{J} \quad , \tag{6-10}$$

the latter being applicable also inside conductors.

6.1.3 Static Conduction Field (DC Conduction Field)

As has been mentioned already in Chapter 1, electric fields with $\sigma \neq 0$ exhibit a conduction flux I in parallel with the electric flux ψ. Because in typical conductors the electric conductivity σ for the conduction flux I significantly exceeds the dielectric conductivity ε for the electric flux ψ, only conduction flux, the familiar current I, is said to exist.

Maintaining a constant conduction flux requires the existence of a dc-source, say a battery, and a closed circuit in which a direct current can flow, Figure 6.2.

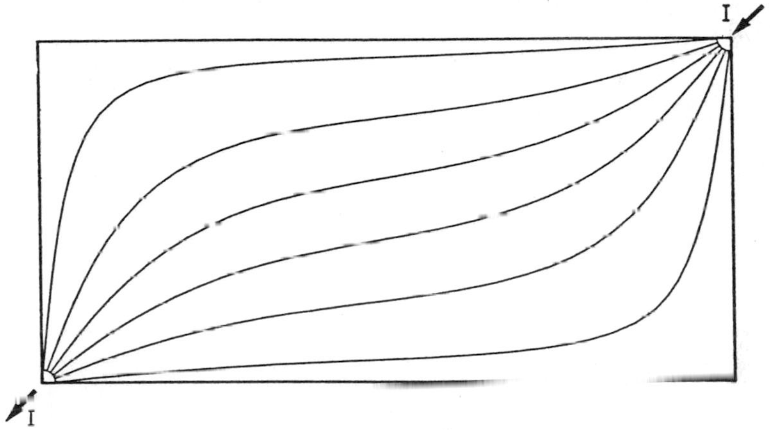

Figure 6.2: Conduction field of a conducting metal sheet diagonally fed with a dc-current.

The vector field $\mathbf{J}_C(\mathbf{r})$ of the current density in conductors is called the *conduction field.* Ohms law in point form applies at each point inside the conductor

$$\boxed{\mathbf{J}_C = \sigma \mathbf{E}} \quad . \tag{6-11}$$

The reader will recognize the formal analogy of this equation with the respective relations between flux densities and field strengths in electrostatic and magnetostatic fields, $\mathbf{D} = \varepsilon\,\mathbf{E}$ and $\mathbf{B} = \mu\,\mathbf{H}$ as in Chapter 1. Depending on the conductivity of a material, different current densities will exist for a given applied field-strength.

As long as the integration path does not include a voltage source, each point of a conduction field obeys the following laws which are also applicable to electrostatic fields,

$$\boxed{\oint \mathbf{E} \cdot d\mathbf{r} = 0} \quad \text{or} \quad \boxed{\text{curl}\,\mathbf{E} = 0} \quad . \tag{6-12}$$

In addition, each region or point external to a voltage source obeys

$$\boxed{\oint \mathbf{J}_C \cdot d\mathbf{S} = 0} \quad \text{or} \quad \boxed{\text{div}\,\mathbf{J}_C = 0} \quad . \tag{6-13}$$

The static or stationary conduction-field is a *vortex-free and source-free* vector field as is the electrostatic field in the environment of electric charges at rest, which obeys

$$\boxed{\oint \mathbf{D} \cdot d\mathbf{S} = 0} \quad \text{or} \quad \boxed{\text{div}\,\mathbf{D} = 0} \quad . \tag{6-14}$$

Due to the formal identity between the flux density $\mathbf{J}_C$ of a current flux I and the flux density $\mathbf{D}$ of an electric flux ψ, electrostatic fields are frequently evaluated via conduction fields on resistive paper or in the electrolytic tank.

Because the static or stationary conduction field is vortex- and source-free, Laplace's potential equation is applicable also to the interior of dc-current carrying conductors.

With $\text{div}\,\mathbf{J}_C = 0$ and $\mathbf{J}_C = \sigma\mathbf{E}$ one obtains upon substitiution

$$\text{div}\,\sigma\mathbf{E} = 0 \quad , \tag{6-15}$$

and with $\mathbf{E} = -\,\text{grad}\,\varphi$

$$\text{div grad}\,\varphi = 0 \quad , \tag{6-16}$$

which, according to Section 4.2, can be written in short form

$$\boxed{\nabla^2\varphi = 0} \quad . \tag{6-17}$$

In other words, all methods generally applied to the solution of electrostatic field problems can be used as well for static conduction fields. Conversely, experimental evaluation of conduction fields can sometimes support an electrostatic field calculation.

As a corollary, for static or stationary conduction fields Maxwell's equations reduce to following equations, Table 6.3.

STATIC CONDUCTION FIELD EQUATIONS	
INTEGRAL FORM	DIFFERENTIAL FORM
$\oint_C \mathbf{E} \cdot d\mathbf{r} = 0$	$\mathrm{curl}\,\mathbf{E} = 0$
$\oint_S \mathbf{J}_C \cdot d\mathbf{S} = 0$	$\mathrm{div}\,\mathbf{J}_C = 0$
$\mathbf{J}_C = \sigma\mathbf{E}$	$\nabla^2\varphi = 0$

Table 6.3: Static conduction field equations

Static or stationary conduction fields are coupled with magnetostatic fields existing inside and outside of the conductors. Because of

$$\mathbf{J}_C = \sigma\,\mathbf{E} \qquad \text{and} \qquad \mathrm{curl}\,\mathbf{H} = \mathbf{J}_C\,, \tag{6-18}$$

this phenomenon is sometimes interpreted as a coupling between stationary electric and magnetic fields. However, "coupling" in stationary fields acts only in one direction. An electric field may be coupled with a magnetic field resulting from the current driven by the electric field, but not the other way round. Depending of the choice of σ, electric and magnetic field strengths may be controlled independently. The electric field can be calculated exclusively from the charge distribution, not knowing anything about the magnetostatic field. Other magnetostatic fields of nearby current-carrying conductors do not interfere with the electric field distribution. In contrast, the nature of the coupling generally used to distinguish between stationary and nonstationary fields is understood as a *mutual* interaction between time-varying electric and magnetic fields. It refers to a reaction via induction of a magnetic field on the electric field which drives the current responsible for that magnetic field (see 6.2).

6.2 Quasi-Stationary Fields

6.2.1 Quasi-Static Electric Fields

Typical examples for quasi-static electric fields are the fields in capacitors, the electric near-field in the close vicinity of antennas, fields between conductors of high-voltage transmission lines, and 60-Hz stray fields of high-voltage components. Whether or not an electric field may be called quasi-static depends on the arrangement considered and the rate-of-change or frequency of the time-varying voltage producing the field.

For instance, between the spheres of a spark gap connected to a 60-Hz ac voltage source there exists at any instant a *source field* $\mathbf{E}_S(\mathbf{r})$ proportional to the respective instantaneous value of the applied voltage. In addition, there flows an alternating current (displacement-current density $\dot{\mathbf{D}}$) whose time-varying magnetic field $\dot{\mathbf{B}}$ induces an electric vortex field $\mathbf{E}_V(\mathbf{r})$. Its vortex density is

$$\text{curl}\ \mathbf{E}_V = -\frac{\partial \mathbf{B}}{\partial t} \tag{6-19}$$

(induction law in differential form, see 3.2.1), Figure 6.3.

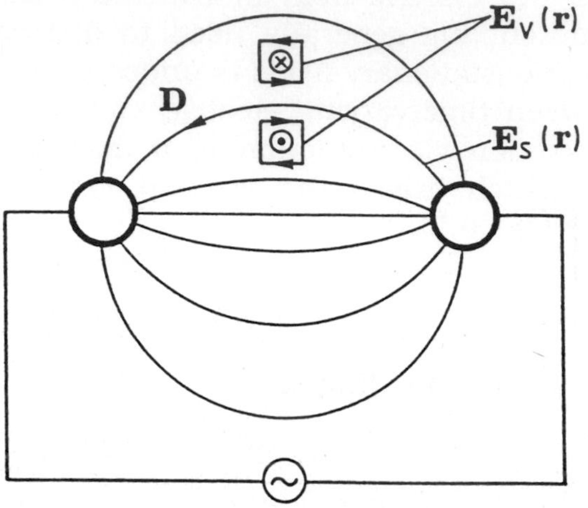

Figure 6.3: Electric field in a spark gap.

In order to allow one to solve for the field strength $\mathbf{E}_V(\mathbf{r})$, the displacement current's magnetic field is derived from its vector magnetic potential, i.e. $\mathbf{B}$ = curl $\mathbf{A}$ (see 5.3), and substituted in the equation above. Interchanging the sequence of differentiation yields

$$\text{curl}\ \mathbf{E}_V = -\frac{\partial \mathbf{B}}{\partial t} = -\frac{\partial}{\partial t}\text{curl}\ \mathbf{A} = -\ \text{curl}\ \frac{\partial \mathbf{A}}{\partial t}\ . \qquad (6\text{-}20)$$

Upon integration (undoing the curl operation $\{\text{curl}\}^{-1}$), we finally obtain the electric vortex-field-strength

$$\boxed{\mathbf{E}_V = -\frac{\partial \mathbf{A}}{\partial t}} \qquad (6\text{-}21)$$

(Regarding uniqueness we refer to Section 5.3).

In arbitrary time-varying electric fields the electric field strength at a field point is generally composed of two components, a *source-* and a *vortex*-field-strength

$$\mathbf{E}(\mathbf{r}) = \mathbf{E}_S(\mathbf{r}) + \mathbf{E}_V(\mathbf{r}) \quad , \tag{6-22}$$

or

$$\boxed{\mathbf{E}(\mathbf{r},t) = -\,\mathrm{grad}\ \varphi(\mathbf{r},t) - \frac{\partial \mathbf{A}(\mathbf{r},t)}{\partial t}} \quad . \tag{6-23}$$

The importance of the second component depends on the rate at which the magnetic vector-potential is varying. At 60 Hz the time rate-of-change of the displacement current's magnetic field is so small that the local contribution of the *vortex-field-strength* $\mathbf{E}_V(\mathbf{r})$ is negligible compared with the *source-field-strength* $\mathbf{E}_S(\mathbf{r})$.Hence, a snap-shot view of a slowly varying field in a capacitor exhibits the same electric field distribution $\mathbf{E}(\mathbf{r})$ as would the electrostatic field produced by a dc-source with a voltage equal to the respective instantaneous value of the ac source. Hence, the quasi-static electric field $\mathbf{E}(\mathbf{r})$ is calculated, as in electrostatic fields, solely from the charge distribution or its respective potential function

$$\mathbf{E}(\mathbf{r},t) = -\,\mathrm{grad}\ \varphi(\mathbf{r},t) \quad . \tag{6-24}$$

All equations given for electrostatic fields apply (see 6.1)

Because quasi-static electric fields typically occur in capacitive arrangements, they are frequently called *capacitive fields* or *high-impedance-fields.*

Quasi-static theory of slowly varying electric fields presumes that a step-change, for instance of the potential of one of the spheres, can be felt simultaneously everywhere in the entire field region. However, because electromagnetic effects can act at a remote point only after a delay time l/v (l: length, distance; v: velocity of light in the medium considered) this presumption holds only if ramp-like changes have a rise-time T_a which can be considered large compared with the propagation time within the arrangement. Hence, the following criteria determine the validity of the quasi-static concept:

Time-Domain : $T_a >> l/v$

Frequency-Domain : $\lambda >> l$

In many cases a risetime of 5 l/v and above (corresponding to $\lambda \geq 5l$) proves to be sufficient.

The reasoning for the frequency-domain criterion will be given during the discussion on electrically short and long transmission lines in Chapter 7.

It is to be noted that l refers to the linear dimension of the field-producing conductor arrangement (near field), not to the dimensions of an object brought into the field. Even a very small electromagnetic shield in the far field of an antenna requires consideration of the field's wave nature; this shield cannot be treated as a quasi-static problem.

6.2.2 Quasi-Static Magnetic Fields

Typical examples of quasi-static magnetic fields are the fields in the windings of rotating electrical machines, those in transformers and reactors, the magnetic near field in the close vicinity of antennas, etc. Whether or not a magnetic field may be called quasistatic depends on the arrangement considered and the time rate-of-change or frequency of the current flowing, or the time-varying voltage driving it. Just as the displacement current in conductors is always neglected, compared to the conduction current (see 3.1.2), so is the magnetic field inside and outside windings exclusively determined by the conduction current. A snap-shot view of a slowly varying magnetic field of a winding exhibits the same field distribution as would the magnetic field caused by a direct current with magnitude equal to the respective instantaneous value of the winding current. Therefore, the quasi-static magnetic field is calculated, like a magnetostatic field, from the instantaneous current distribution of the conduction current only, or its corresponding potential function

$$\boxed{\mathbf{H}(\mathbf{r}) = \text{grad}\ \varphi_m(\mathbf{r})} \quad \text{or} \quad \boxed{\mathbf{B}(\mathbf{r}) = \text{curl}\ \mathbf{A}(\mathbf{r})} \ . \qquad (6\text{-}25)$$

The equations given for the magnetostatic field (see 6.1.2) are generally applicable.

Because quasi-static magnetic fields typically occur in inductive arrangements, they are frequently called *inductive fields* and, because of the generally low impedance of the circuit, are often referred to as *low-impedance fields.*

In windings with several turns the stray capacitance between individual turns allows a considerable displacement current; none the less, the electric field of such a stray capacitance is still quasi-static. Also, *self-induction* is a quasi-static or quasi-stationary effect.

A winding looses its quasi-stationary quality if its total wire length, or the traveling time l/v calculated from it, is no longer small compared with the risetime T_a of a ramp change of the applied voltage. The same criteria, as derived for the quasi-static electric field, apply

$$\text{Time-Domain: } T_a >> l/v$$

$$\text{Frequency-Domain: } \lambda \gg l \,.$$

If these criteria are complied with, a winding is considered *electrically short* or called a *lumped element*. If not, it need be treated as an *electrically long* line with distributed parameters (transmission line). Under certain conditions, quasi-stationary methods may be applied also to electrically long lines (traveling wave theory, see Chapter 7). The problem *electrically long* or *electrically short* occurs also with capacitors whose high capacitance is due to size.

6.2.3 Quasi-Static Conduction Fields

Quasi-static conduction fields exist in conductors carrying ac, in current-viewing resistors, in electrolytic tanks etc., presuming that the skin effect is not substantial enough. Whether a conduction field may be called quasi-static or not depends again on the arrangement considered and the time rate-of-change or frequency of the time-varying voltage applied. Since we are only dealing with the interior of conductors, the displacement current can essentially be neglected. The conduction field is then said to be quasi-static only if the vortex strength

$$\oint \mathbf{E} \cdot d\mathbf{r} \tag{6-26}$$

or the vortex density curl $\mathbf{E}$ of the electric vortex field resulting from the conduction current's magnetic field are negligible in magnitude. Since a snap-shot view of $\mathbf{J}_C(\mathbf{r},t)$ of a time-varying conduction field coincides with the static conduction field of a respective direct current, all equations for the static conduction field apply to the quasi-static conduction field as well. Regarding the entire conductor loop,

$$\oint \mathbf{E} \cdot d\mathbf{r} \tag{6-27}$$

may differ from zero (self-induction law, see Section 3.5).

6.2.4 Conduction Fields with Skin Effect

With higher time rates-of-change or higher frequencies, the electric vortex field induced in the conductor by the conduction current's magnetic field can no longer be neglected compared to the applied electric field. Both electric fields are oriented such that they reinforce each other at the conductor surface whereas they oppose each other in the conductor center axis, leading to the skin effect at higher frequencies, Figure 6.4.

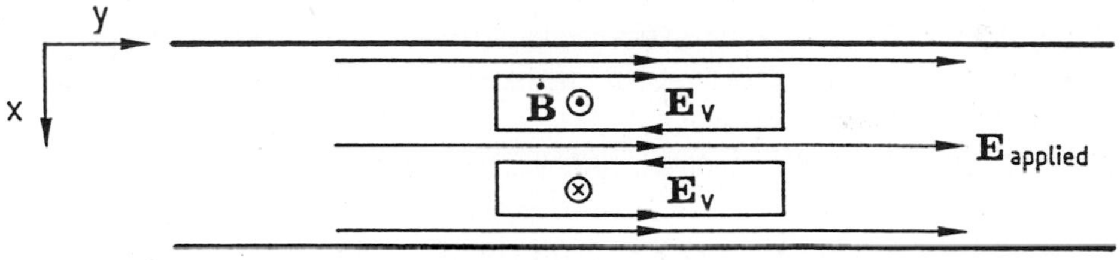

Figure 6.4: Longitudinal cross-section of a circular conductor with skin effect.

The conduction field no longer obeys Laplace's equation $\nabla^2\varphi = 0$. Rather, the fields $\mathbf{E}$, $\mathbf{J}_c$, $\mathbf{H}$, $\mathbf{B}$ are now functions of space and time. The electric field strength in conductors must satisfy two conditions. In the first one, the displacement current is neglected compared to the conduction current (see 3.2)

$$\text{curl}\,\mathbf{H} = \sigma\,\mathbf{E} + \cancel{\varepsilon\frac{\partial \mathbf{E}}{\partial t}} \quad \text{and} \quad \text{curl}\,\mathbf{E} = -\frac{\partial \mathbf{B}}{\partial t} = -\mu\frac{\partial \mathbf{H}}{\partial t} \,. \tag{6-28}$$

From the pair of equations for $\mathbf{E}$ and $\mathbf{H}$ we eliminate $\mathbf{H}$ by applying the differential operator curl to both sides of the right equation and by interchanging the order of sequence of the differentiation with respect to time and space

$$\text{curl curl } \mathbf{E} = -\mu \text{ curl } \frac{\partial \mathbf{H}}{\partial t} = -\mu \frac{\partial}{\partial t} \text{curl } \mathbf{H} \ . \tag{6-29}$$

Substituting $\sigma\mathbf{E}$ for curl $\mathbf{H}$ yields

$$\text{curl curl } \mathbf{E} = -\sigma\mu \frac{\partial \mathbf{E}}{\partial t} \ . \tag{6-30}$$

Employing again the vector identity encountered already in Section 5.4 (see also A3),

$$\text{curl curl } \mathbf{X} = \text{grad div } \mathbf{X} - \nabla^2 \mathbf{X} \ , \tag{6-31}$$

we obtain

$$\text{grad div } \mathbf{E} - \nabla^2 \mathbf{E} = -\sigma\mu \frac{\partial \mathbf{E}}{\partial t} \ , \tag{6-32}$$

Since the conduction field is source-free, i.e.,

$$\text{div } \mathbf{J}_C = \text{div } \sigma\mathbf{E} = 0 \ , \tag{6-33}$$

grad div $\mathbf{E}$ vanishes as well. Thus we arrive at the *diffusion* equation (see 8.4)

$$\boxed{\nabla^2 \mathbf{E} = \sigma\mu \frac{\partial \mathbf{E}}{\partial t}} \ . \tag{6-34}$$

To facilitate the mathematical treatment of numerous field problems involving sinusoidal, time-harmonic quantities, this equation is frequently used in its complex notation (see A4)

$$\boxed{\nabla^2 \underline{\mathbf{E}} = j\omega\sigma\mu\, \underline{\mathbf{E}}} \quad . \tag{6-35}$$

Eliminating the electric field strength **E** in the previous pair of equations instead of the magnetic field strength **H** yields for the latter

$$\boxed{\nabla^2 \mathbf{H} = \sigma\mu \frac{\partial \mathbf{H}}{\partial t}} \quad \text{or} \quad \boxed{\nabla^2 \underline{\mathbf{H}} = j\omega\sigma\mu \underline{\mathbf{H}}} \quad . \tag{6-36}$$

Finally, with $\mathbf{J}_C = \sigma \mathbf{E}$, a similar equation for the conduction field is obtained

$$\boxed{\nabla^2 \mathbf{J}_C = \sigma\mu \frac{\partial \mathbf{J}_C}{\partial t}} \quad \text{or} \quad \boxed{\nabla^2 \underline{\mathbf{J}}_C = j\omega\sigma\mu \underline{\mathbf{J}}_C} \quad . \tag{6-37}$$

Instead of $j\omega\sigma\mu$, the abbreviation k^2 with $k = \sqrt{j\omega\sigma\mu}$ is often used. With k, the complex equations read as follows

$$\boxed{\nabla^2 \underline{\mathbf{E}} = k^2 \underline{\mathbf{E}}} \quad \text{or} \quad \boxed{\nabla^2 \underline{\mathbf{H}} = k^2 \underline{\mathbf{H}}} \quad . \tag{6-38}$$

For one-dimensional field problems, the solution takes the form

$$\underline{\mathbf{E}}(x) = \underline{\mathbf{E}}(0)\ e^{-kx} \tag{6-39}$$

using the left equation as an example.

Substituting in this equation $k = (1-j)/\delta$ yields the field-strength distribution inside a plane conductor with skin effect (x being measured from the surface)

$$\underline{\mathbf{E}}_y(x) = \underline{\mathbf{E}}_y(0)\ e^{-x(1/\delta + j/\delta)} = \underline{\mathbf{E}}_y(0)\ e^{-x/\delta}\ e^{-jx/\delta}\ , \tag{6-40}$$

The factor $e^{-x/\delta}$ describes the attenuation, the factor $e^{-jx/\delta}$ the phase-shift. At a distance $x=\delta$ below the surface, the field strength´s magnitude has decreased by the factor e

$$|\underline{\mathbf{E}}_y(\delta)| = \underline{\mathbf{E}}_y(0)\ e^{-1}\ . \tag{6-41}$$

Therefore, the quantity δ is called 1/*e-penetration depth* or *skin depth*

$\mu = 4\pi \times 10^{-7}$
$\sigma_{Cu} = 5.80 \times 10^{7}$

$$\delta = \sqrt{\frac{1}{\pi f \sigma \mu}} = \sqrt{\frac{2}{\omega \sigma \mu}}\ . \tag{6-42}$$

For $\delta >> d$, e.g. the largest linear dimension of a conductor cross-section, the skin effect can be neglected. Then, a quasi-static conduction field is said to exist.

The preceding space- and time-dependent differential equations for the electric and magnetic field-strengths do not describe the propagation of a wave in the usual sense but a diffusion process instead. In fact, they are the electric and magnetic analogies of the renowned *diffusion equation*. In contrast to genuine wave processes in which the effect of a local disturbance can be detected at a remote point only after the elapse of a certain delay time (see 6.3.2), in a diffusion process a disturbance is simultaneously felt in the entire system, provided the measuring equipment is sufficiently sensitive. This is what makes the difference between a uniform transmission line (continuum problem) and its ladder-structure equivalent circuit composed of a finite number of identical π sections (lumped components). Regarding diffusion processes, defining a propagation velocity or a delay time does not make much sense, because for $\sigma \neq 0$ the magnitude of the latter would only depend on the detection system's sensitivity. Hence, the conduction field with skin effect, lacking a true propagation quality, does not belong to non-stationary fields, but to the group of quasi-stationary fields. Of course, just as in the magnetostatic field of a coil or a quasi-static electric field in a capacitor, the conductor dimensions must be small compared with the wave-length.

6.3 Non-Stationary Fields - Electromagnetic Waves

6.3.1 Wave Equation

When the risetimes of transients attain the order of magnitude of propagation times or delay times within systems, or when the wave lengths of sinusoidal quantities attain the order of magnitude of the system's linear dimensions, time-varying fields loose their stationary quasi-static nature. The fields detach from the conductors and propagate as electromagnetic waves into space. For instance, a rapidly varying magnetic field $i_0(t)$ in a conductor is linked with a concentric, rapidly varying magnetic field $\mathbf{H}_0(\mathbf{r},t)$ whose field lines, according to the induction law, are surrounded by an electric vortex-field $\mathbf{E}_0(\mathbf{r},t)$. The displacement currents flowing along the electric vortex-field lines are again surrounded by concentric magnetic vortex-field lines. This is schematically depicted in Figure 6.5.

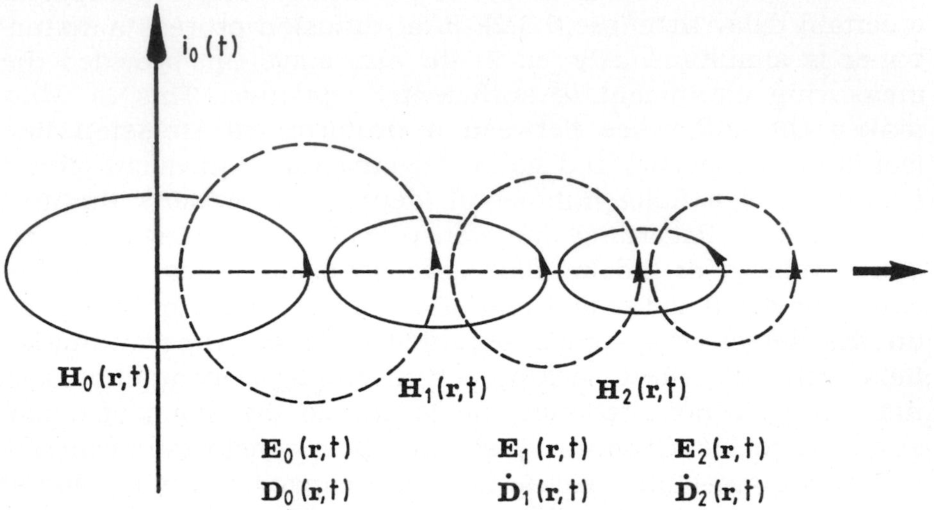

Figure 6.5: Schematic illustration of the propagation process of electromagnetic waves.

Completing this scheme in our imagination for the entire space, all partial fields $\mathbf{H}_\nu(\mathbf{r},t)$ superimpose to form solenoidal magnetic field lines surrounding the conductor concentrically. The propagation ability can be grasped from the fact that because of the vortex nature, each electric field emanating from the previous time-varying magnetic field and each magnetic field emanating from the previous time-varying electric field necessarily possess a wider spatial extension than the respective previous field.

In contrast to all previous considerations in which the displacement current or its magnetic field have always been neglected compared to the conduction current or its magnetic field, in dielectrics and free space we need to neglect the conduction current compared to the displacement current. Then, the displacement current is the only existing current. When dealing with wave problems, we frequently speak of the time-varying flux density $\dot{\mathbf{D}}$ instead of the displacement-current density $\mathbf{J}_d$ (although the concept of displacement current may be helpful in visualizing and in modeling transmission, reception, and propagation problems, see Figure 6.5).

In ideal, loss-free dielectrics (σ=0) the space- and time-dependent field quantities $\mathbf{E}(\mathbf{r},t)$ and $\mathbf{H}(\mathbf{r},t)$ must obey the following two equations:

$$\text{curl}\,\mathbf{H} = \cancel{\sigma\mathbf{E}} + \varepsilon\frac{\partial\mathbf{E}}{\partial t} \quad \text{and} \quad \text{curl}\,\mathbf{E} = -\mu\frac{\partial\mathbf{H}}{\partial t} \quad . \tag{6-43}$$

Here again, we have two equations in two unknowns, $\mathbf{E}$ and $\mathbf{H}$. From these equations we eliminate $\mathbf{H}$ by taking the curl of both sides of the right equation and by interchanging the order of the sequence of the space- and time differentiation,

$$\text{curl curl}\,\mathbf{E} = -\mu\,\text{curl}\frac{\partial\mathbf{H}}{\partial t} = -\mu\frac{\partial}{\partial t}\text{curl}\,\mathbf{H} \quad . \tag{6-44}$$

Substituting $\varepsilon(\partial\mathbf{E}/\partial t)$ for curl $\mathbf{H}$ we obtain

$$\text{curl curl}\,\mathbf{E} = -\varepsilon\mu\frac{\partial^2\mathbf{E}}{\partial t^2} \quad . \tag{6-45}$$

Finally we apply again the vector identity encountered in Section 5.4 (see also A3)

$$\text{curl curl}\,\mathbf{X} = \text{grad div}\,\mathbf{X} - \nabla^2\mathbf{X} \quad , \tag{6-46}$$

and obtain

$$\text{grad div}\,\mathbf{E} - \nabla^2\mathbf{E} = -\varepsilon\mu\frac{\partial^2\mathbf{E}}{\partial t^2} \quad . \tag{6-47}$$

In free space (without space-charge density ρ) we have div **E** = 0, hence, grad div **E** = 0. Thus we obtain the wave equation for the electric field strength

$$\nabla^2 \mathbf{E} - \varepsilon\mu \frac{\partial^2 \mathbf{E}}{\partial t^2} = 0 \quad . \tag{6-48}$$

Analogously we obtain by elimination of **E** the wave equation for the magnetic field strength **H**

$$\nabla^2 \mathbf{H} - \varepsilon\mu \frac{\partial^2 \mathbf{H}}{\partial t^2} = 0 \quad . \tag{6-49}$$

Both vector differential equations can be decomposed into three scalar wave equations

$$\nabla^2 E_x - \mu\varepsilon \frac{\partial^2 E_x}{\partial t^2} = 0 \qquad \nabla^2 H_x - \mu\varepsilon \frac{\partial^2 H_x}{\partial t^2} = 0$$

$$\nabla^2 E_y - \mu\varepsilon \frac{\partial^2 E_y}{\partial t^2} = 0 \qquad \nabla^2 H_y - \mu\varepsilon \frac{\partial^2 H_y}{\partial t^2} = 0$$

$$\nabla^2 E_z - \mu\varepsilon \frac{\partial^2 E_z}{\partial t^2} = 0 \qquad \nabla^2 H_z - \mu\varepsilon \frac{\partial^2 H_z}{\partial t^2} = 0 \quad . \tag{6-50}$$

To facilitate mathematical treatment, the wave equations are frequently used in their complex notation (see A5)

$$\nabla^2 \underline{\mathbf{E}} - (j\omega)^2 \, \varepsilon\mu \, \underline{\mathbf{E}} = 0 \qquad \nabla^2 \underline{\mathbf{H}} - (j\omega)^2 \, \varepsilon\mu \, \underline{\mathbf{H}} = 0 \tag{6-51}$$

or

$$\nabla^2 \underline{\mathbf{E}} + \omega^2 \, \varepsilon\mu \, \underline{\mathbf{E}} = 0 \qquad \nabla^2 \underline{\mathbf{H}} + \omega^2 \, \varepsilon\mu \, \underline{\mathbf{H}} = 0 \tag{6-52}$$

More compactly, with $k_0 = \sqrt{\varepsilon\mu}$

$$\nabla^2 \underline{\mathbf{E}} + k_0^2 \, \underline{\mathbf{E}} = 0 \qquad \nabla^2 \underline{\mathbf{H}} + k_0^2 \, \underline{\mathbf{H}} = 0 \tag{6-53}$$

Evaluation of the electric and magnetic field-strength of a wave process in free space for certain boundary conditions can be accomplished either directly by solving the previous wave equations, or indirectly by means of so-called retarded potentials, which shall be introduced in the next section. Frequently one is content with the investigation of the propagation of TEM waves (electromagnetic waves with transverse **E** and **H** field vectors), for which the system of equations is greatly simplified, presuming proper choice is made of the coordinate system. Solution of the wave equations in time or frequency domain is beyond the scope of this introductory text of field theory concepts.

6.3.2 Retarded Potentials

In much the same way that the evaluation of the field quantities **E** and **H** of static and quasi-static fields is facilitated by employing potential functions, the mathematical treatment of wave problems can be simplified by so called *retarded potentials*. In contrast to static and quasi-static fields in which a disturbance of a given charge aggregate can be simultaneously felt in the entire field region, in spatially extended regions a disturbance of the source distribution can be felt at a remote field point only after the elapse of a propagation time l/v, where

$$v = \frac{1}{\sqrt{\varepsilon\mu}} \tag{6-54}$$

is the appropriate velocity of light in the medium. Particularly with distributed sources, the individual contributions of local changes will arrive at a particular field point after different propagation time, resulting in different net potentials, compared with the static or quasi-static case. Retarded potentials are solutions of wave equations for electromagnetic potentials which will be derived below.

It has beeen shown in Sections 5.3 (magnetic vector potential) and 6.2.1 (quasi-static electric field) that the field strengths **H** and **E** can be represented as functions of potentials $\varphi(\mathbf{r})$ and $\mathbf{A}(\mathbf{r})$

$$\boxed{\mathbf{H} = \frac{1}{\mu}\,\text{curl}\,\mathbf{A}} \quad \text{and} \quad \boxed{\mathbf{E} = -\,\text{grad}\,\varphi - \frac{\partial \mathbf{A}}{\partial t}}\;. \tag{6-55}$$

Substituting the right sides of these equations into Maxwell´s equations for the field strengths **E** and **H** we obtain the wave equations for the potentials. First we substitute the magnetic flux density **B** by **A** in Ampere´s circuital law

$$\text{curl}\,\mathbf{H} = \mathbf{J} = \mathbf{J}_C + \varepsilon\frac{\partial \mathbf{E}}{\partial t}\;, \tag{6-56}$$

$$\text{curl } \mathbf{H} = \frac{1}{\mu} \text{curl } \mathbf{B} = \frac{1}{\mu} \text{curl curl } \mathbf{A} = \mathbf{J}_C + \varepsilon \frac{\partial \mathbf{E}}{\partial t} \quad . \qquad (6\text{-}57)$$

Then we replace in the latter equation **E** by - grad $\varphi - \partial\mathbf{A}/\partial t$ (see 6.2.1)

$$\frac{1}{\mu} \text{curl curl } \mathbf{A} = \mathbf{J}_C - \varepsilon \frac{\partial}{\partial t}\left(\text{grad } \varphi + \frac{\partial \mathbf{A}}{\partial t}\right) = \mathbf{J}_C - \varepsilon \text{ grad } \frac{\partial \varphi}{\partial t} - \varepsilon \frac{\partial^2 \mathbf{A}}{\partial t^2},$$

$$\text{curl curl } \mathbf{A} = \mu \mathbf{J}_C - \varepsilon\mu \text{ grad } \frac{\partial \varphi}{\partial t} - \varepsilon\mu \frac{\partial^2 \mathbf{A}}{\partial t^2}. \qquad (6\text{-}58)$$

Here, the reader correctly expects to once again modify the term curl curl **A** by the now familiar vector identity

$$\text{curl curl } \mathbf{X} = \text{grad div } \mathbf{X} - \nabla^2 \mathbf{X} \qquad (6\text{-}59)$$

so that

$$\text{grad div } \mathbf{A} - \nabla^2 \mathbf{A} = \mu \mathbf{J}_C - \varepsilon\mu \text{ grad } \frac{\partial \varphi}{\partial t} - \varepsilon\mu \frac{\partial^2 \mathbf{A}}{\partial t^2} \quad . \qquad (6\text{-}60)$$

Factoring out the gradient operation yields

$$\nabla^2 \mathbf{A} - \varepsilon\mu \frac{\partial^2 \mathbf{A}}{\partial t^2} = -\mu \mathbf{J}_C + \text{grad}\left(\text{div } \mathbf{A} + \varepsilon\mu \frac{\partial \varphi}{\partial t}\right) \quad . \qquad (6\text{-}61)$$

In Section 5.3 on the magnetic vector-potential we have learned that a certain vector field of the magnetic flux density **B**(**r**) may have various vector potentials which may differ in their divergence, because the existence of a source-field is no longer evident upon application of the curl operation. Therefore, we are free to choose for div **A** a function that fits our purpose best and we decide in favour of

$$\boxed{\operatorname{div} \mathbf{A} = - \varepsilon\mu \frac{\partial \varphi}{\partial t}} \quad . \tag{6-62}$$

Introducing this function into the earlier equation causes the second term on its right side to vanish. This convention is called *Lorentz condition* or *Lorentz gauge* for the potential (see 4.5). At this point we remember the convention div **A** = 0, agreed upon in magnetostatic fields, referred to as the *Coulomb gauge* (see 5.4).

Thus, we obtain the inhomogeneous wave equation for the magnetic vector-potential

$$\boxed{\nabla^2 \mathbf{A} - \varepsilon\mu \frac{\partial^2 \mathbf{A}}{\partial t^2} = - \mu \mathbf{J}_C} \quad . \tag{6-63}$$

The inhomogeneous wave equation for the electric scalar-potential is obtained upon substituting

$$\mathbf{E} = - \operatorname{grad} \varphi - \partial \mathbf{A}/\partial t \tag{6-64}$$

into the differential form of Gauss´s law for the electric field and interchanging the order of the sequence of the differential operations ∂/∂t and div.

$$\operatorname{div} \mathbf{D} = \operatorname{div} \varepsilon\mathbf{E} = \operatorname{div} \varepsilon \left(-\operatorname{grad} \varphi - \frac{\partial \mathbf{A}}{\partial t}\right) = \rho \tag{6-65}$$

$$\text{div grad } \varphi + \frac{\partial}{\partial t} \text{div } \mathbf{A} = -\frac{\rho}{\varepsilon} \quad . \tag{6-66}$$

Substituting again div **A** by means of the Lorentz condition we obtain

$$\boxed{\nabla^2\varphi - \varepsilon\mu \frac{\partial^2\varphi}{\partial t^2} = -\frac{\rho}{\varepsilon}} \quad . \tag{6-67}$$

Apparently, by introducing the Lorentz condition, the electric and magnetic potentials become uncoupled. However, this uncoupling must not be misunderstood in the sense that ρ and $\mathbf{J}_C$ may be independently chosen. Only pairs of ρ and $\mathbf{J}_C$ which simultaneously obey the law of continuity

$$\text{div } \mathbf{J}_C = -d\rho/dt \tag{6-68}$$

or

$$\oint \mathbf{J}_C \cdot d\mathbf{S} = -\frac{dQ}{dt} \tag{6-69}$$

(s.3.2.5. and 3.1.5), satisfy as waves the wave equation.

Solving the two wave equations for the potentials, given current and charge distributions, yields the retarded potentials

$$\mathbf{A}(\mathbf{r},t) = \int \frac{\mu\, \mathbf{J}_C\left(\mathbf{r}_0, \left(t - \frac{|\mathbf{r}-\mathbf{r}_0|}{v}\right)\right)}{4\pi |\mathbf{r}-\mathbf{r}_0|}\, dV$$

$$\varphi(\mathbf{r},t) = \int \frac{\rho\left(\mathbf{r}_0, \left(t - \frac{|\mathbf{r}-\mathbf{r}_0|}{v}\right)\right)}{4\pi\varepsilon |\mathbf{r}-\mathbf{r}_0|}\, dV \quad . \qquad (6\text{-}70)$$

These equations ought to look familiar to the reader, differing from the ordinary potentials of Sections 4.3 and 5.3 merely by their retardation (delay) $|\mathbf{r} - \mathbf{r}_0|/v$. The retardation vanishes for $|\mathbf{r} - \mathbf{r}_0|/v \ll t$, i.e. for the quasi-static case.

6.3.3 Hertz Potentials

In the previous section we have expressed the retarded potentials $\varphi(\mathbf{r},t)$ and $\mathbf{A}(\mathbf{r},t)$ in terms of given charge and current densities $\rho(\mathbf{r},t)$ and $\mathbf{J}_C(\mathbf{r},t)$. In addition to that, it is possible to express the potentials in terms of a generalized *electric polarization*

$$\mathbf{P}(\mathbf{r},t) = \mathbf{D} - \varepsilon_0\, \mathbf{E} \qquad (6\text{-}71)$$

and *magnetization*

$$\mathbf{M}(\mathbf{r},t) = \frac{\mathbf{B}}{\mu_0} - \mathbf{H} \ . \qquad (6\text{-}72)$$

Usually, these quantities describe material properties, yet they can be interpreted as ultimate macroscopic manifestations of the effects of atomic charges and current densities.

Let us assume a source- and vortex-free field region with $\rho=0$, $\mathbf{J}_C=0$, i.e. no free charges and currents. If we rewrite the electric flux-density **D** in Maxwell's equations in terms of the polarization **P**, and the magnetic field-strength **H** in terms of the magnetization **M** we find a new system of equations in which we can formally assign a "charge density" $\rho'(\mathbf{r})$ to the polarization and a "current density" to a term composed of polarization and magnetization $\mathbf{J}'(\mathbf{r})$,

$$\rho' = -\operatorname{div}\mathbf{P} \qquad \text{and} \qquad \mathbf{J}_C' = \frac{\partial \mathbf{P}}{\partial t} + \operatorname{curl}\mathbf{M} \tag{6-73}$$

Recalling the previous Section 6.3.2, these relations can likewise be interpreted as excitations of retarded potentials in free space

$$\mathbf{A}(\mathbf{r},t) = \int_V \frac{\mu\left(\frac{\partial \mathbf{P}(\mathbf{r}_0,t')}{\partial t} + \operatorname{curl}\mathbf{M}(\mathbf{r}_0,t')\right)}{4\pi\,|\mathbf{r}-\mathbf{r}_0|}\,dV\;, \tag{6-74}$$

$$\varphi(\mathbf{r},t) = -\int_V \frac{\operatorname{div}\mathbf{P}(\mathbf{r}_0,t')}{4\pi\varepsilon\,|\mathbf{r}-\mathbf{r}_0|}\,dV \tag{6-75}$$

where $t' = \frac{t - |\mathbf{r}-\mathbf{r}_0|}{c}$ represents the retarded time (s.6.3.2).

These equations look by no means more inviting for a solution than the initial potential equations. Therefore, Hertz introduced two new vector potentials Π_e and Π_m related to the conventional potentials by

$$\varphi = -\operatorname{div}\Pi_e \qquad \text{and} \qquad \mathbf{A} = \frac{\partial \Pi_e}{\partial t} + \operatorname{curl}\Pi_m\;. \tag{6-76}$$

It is to be noted that these relations formally coincide with the relations between the equivalent charge- and current density distributions ρ' and $\mathbf{J}_C'$ and their corresponding generalized polarization and magnetization introduced at the beginning of this section.

Substituting for φ and **A** in the potential's wave equations (see 6.3.2) the right sides of the equations above, one obtains new wave equations in the *Hertz potentials* Π_e and Π_m with **P** and **M** as excitations for forcing functions,

$$\nabla^2 \Pi_e - \varepsilon\mu \frac{\partial^2 \Pi_e}{\partial t^2} = -\mathbf{P}$$

$$\nabla^2 \Pi_m - \varepsilon\mu \frac{\partial^2 \Pi_m}{\partial t^2} = -\mathbf{M} \qquad (6\text{-}77)$$

As was already shown in section 6.3.2, such equations have the solutions

$$\Pi_e(\mathbf{r},t) = \int_V \frac{\mathbf{P}(\mathbf{r}_0, t')}{4\pi\varepsilon\, |\mathbf{r} - \mathbf{r}_0|}\, dV$$

$$\Pi_m(\mathbf{r},t) = \int_V \frac{\mu\mathbf{M}(\mathbf{r}_0, t')}{4\pi\, |\mathbf{r} - \mathbf{r}_0|}\, dV \qquad (6\text{-}78)$$

which are obviously simpler than the solution $\varphi(\mathbf{P})$ and $\mathbf{A}(\mathbf{P},\mathbf{M})$. Introducing the Hertz potentials into the equations for the conventional potentials φ and **A** we obtain

$$\mathbf{E} = - \operatorname{grad} \varphi - \frac{\partial \mathbf{A}}{\partial t} \quad , \tag{6-79}$$

and

$$\mathbf{H} = \frac{1}{\mu} \operatorname{curl} \mathbf{A} \quad , \tag{6-80}$$

which yields, upon applying strictly differential operations, the desired field vectors **E**, **D**, **H**, **B**.

6.3.4 Energy Density of Electric and Magnetic Fields Energy-Flux Density of Electromagnetic Waves

The propagation of electromagnetic waves is associated with the transportation of energy composed of contributions from the electric and magnetic fields. When we try to separate the individual components and describe their distribution in space we encounter the same difficulties as with fluxes, vortex strengths and source strengths. This is due to the fact that the energy of a field point, having zero volume, is identically zero (see Chapter 1 and 3). To overcome this problem, we form the ratio of energy to volume, whose limit remains finite even as the volume approaches zero, yielding energy densities of electric and magnetic fields

$$w_e(\mathbf{r},t) = \frac{1}{2} \varepsilon \, \mathbf{E}^2(\mathbf{r},t) \quad \text{and} \quad w_m(\mathbf{r},t) = \frac{1}{2} \mu \, \mathbf{H}^2(\mathbf{r},t) \; . \tag{6-81}$$

To obtain the energy of a particular field region for a given field-strength distribution, we multiply the densities by the volume considered (in uniform fields) or evaluate the integrals

$$W_e = \int_V \frac{1}{2} \varepsilon \mathbf{E}^2 \, dV \quad \text{and} \quad W_m = \int_V \frac{1}{2} \mu \mathbf{H}^2 \, dV \quad . \tag{6-82}$$

The mutual coupling of electric and magnetic energy in an electromagnetic wave results in a *directed energy transport*, or an *energy flux*. An energy flux has the dimensions of energy per time, hence it represents *power*. In Chapter 1 it has been shown, that a flux always implies the existence of an area through which it flows. In order to allow an exclusive statement about the electromagnetic field property at a point one defines an *energy-flux density* or *power density*. The energy-flux density or power density in the field of an electromagnetic wave is given by the Poynting vector **S**,

$$\mathbf{S} = \mathbf{E} \times \mathbf{H} \quad . \tag{6-83}$$

Regarding its physical nature, the Pointing vector is a density function of position in much the same way as are electric flux density, vortex density, and source density. It describes the direction and magnitude of the energy flux at discrete field points. The Poynting vector has the dimensions of Watt/m^2. Energy flux-density and power density are interchangeably used.

7 Transmission-Line Equations

The reader might wonder why in *field theory concepts* we suddenly deal with transmission lines which actually go with circuit theory concepts. True enough, but we need to realize that circuit theory relations are only special cases of more general Maxwell equations of field theory. Basically, all engineering problems can be solved employing field theory equa tions. Unfortunately, during the solution process of these differential equations (integration) there occur integration constants or functions (as with indefinite integrals) which must be determined from problem-specific boundary conditions in a generally cumbersome manner. Hence, whenever possible, one uses network theory which is much easier to handle. It is interesting to note, that while developing the mathematics for a circuit model of transmission lines (before the discovery of the real existence of electromagnetic waves by Heinrich Hertz 1887-1889 at the University of Karlsruhe) Oliver Heaviside found a differential equation which later proved to be the mother of numerous other differential equations of field theory, indeed the entire field of mathematical physics, i.e. the *telegraphist's equation.* This equation for currents and voltages of electrically long lines is encountered in the time domain - v(x,t), i(x,t) - and in the frequency domain $\underline{V}(x)$, $\underline{I}(x)$.

We derive in this chapter the telegraphist's equation using familiar circuit analysis. This serves only to provide the reader with a very descriptive basis for digesting the typical differential equations of field theory and mathematical physics presented in the next chapter.

We begin with an electrically long homogeneous line, that is a line with constant electric and magnetic parameters in the propagation direction, Figure 7.1.

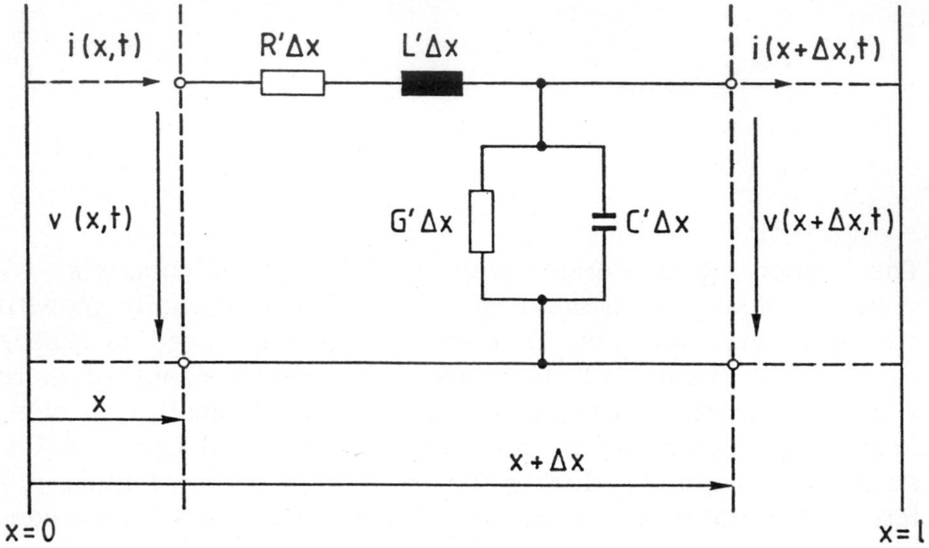

Figure 7.1: Equivalent circuit of a short section Δx of an electrically long homogeneous line.

The resistance R and the inductance L of the current loop formed by both conductors, as well as the capacitance C and the conductance G between both conductors, are expressed in per-unit length

R'	$= \Delta R/\Delta l$	resistance per unit length
L'	$= \Delta L/\Delta l$	inductance per unit length
C'	$= \Delta C/\Delta l$	capacitance per unit length
G'	$= \Delta G/\Delta l$	conductance per unit length

The reader should be aware that R' and L' are derived for both conductors in series.

The axial coordinate x may be measured either from the input or output end of the line. The first case is found useful for the discussion of time-harmonic voltages and currents or from the point of view of a load at the line´s end, the latter if the propagation of traveling waves is to be considered. We choose x=0 at the line´s input, Figure 7.1.

Let the incremental section Δx be electrically short, then the time-dependent quantities v(t) and i(t) lack a space-dependency. Hence, their values are only determined by the lumped elements of this sections´s equivalent circuit, allowing quasi-static treatment by means of Kirchhoff´s laws.

VOLTAGE LAW $\Sigma V = 0$:

$$R'\Delta x\, i(x,t) + L'\Delta x \frac{di(x,t)}{dt} + v(x+\Delta x,t) - v(x,t) = 0 \quad .$$

Dividing by Δx, we obtain

$$R'i(x,t) + L' \frac{di(x,t)}{dt} + \frac{v(x+\Delta x,t) - v(x,t)}{\Delta x} = 0 \quad .$$

Taking the limit as Δx approaches zero gives

$$\Delta x \rightarrow 0 : \; R'i(x,t) + L' \frac{\partial i(x,t)}{\partial t} + \frac{\partial v(x,t)}{\partial x} = 0 \quad ,$$

$$\boxed{R'i(x,t) + L' \frac{\partial i(x,t)}{\partial t} = \frac{-\partial v(x,t)}{\partial x}} \quad . \qquad (7\text{-}1)$$

CURRENT LAW $\sum I = 0$:

$$i(x,t) - i(x+\Delta x,t) - G'\Delta x\, v(x+\Delta x,t) - C'\Delta x \frac{dv(x+\Delta x,t)}{dt} = 0 \ .$$

Dividing by Δx, we obtain

$$\frac{i(x,t) - i(x+\Delta x,t)}{\Delta x} - G' v(x+\Delta x,t) - C' \frac{dv(x+\Delta x,t)}{dt} = 0 \ .$$

Taking the limit as Δx approaches zero gives

$$\Delta x \to 0 : \qquad -\frac{\partial i(x,t)}{\partial x} = G'\, v(x,t) + C' \frac{\partial v(x,t)}{\partial t} \ ,$$

$$\boxed{G' v(x,t) + C' \frac{\partial v(x,t)}{\partial t} = -\frac{\partial i(x,t)}{\partial x}} \ . \qquad (7\text{-}2)$$

Thus, we have obtained two equations for the two unknowns v(x,t) and i(x,t). Differentiating one equation with respect to x, the other with respect to t and substituting one into the other will result in the elimination of v and i. Hence, we obtain the complete transmission-line equations in the time-domain,

$$\boxed{\begin{aligned} \frac{\partial^2 v}{\partial x^2} &= L'C' \frac{\partial^2 v}{\partial x^2} + (R'C' + L'G') \frac{\partial v}{\partial t} + R'G'v \\ \frac{\partial^2 i}{\partial x^2} &= L'C' \frac{\partial^2 i}{\partial x^2} + (R'C' + L'G') \frac{\partial i}{\partial t} + R'G'i \end{aligned}} \ . \qquad (7\text{-}3)$$

These equations allow the calculation of v(x,t) and i(x,t), given specific initial conditions or boundary values. The solution in the time-domain represents the basis of the powerful traveling wave theory. Restricting ourselves to sinusoidal excitation (frequency domain) allows complex notation of the equations, where by the partial differential equations simplify to ordinary differential equations. Then, the time-varying voltages v(t) and currents i(t) become phasors (i.e. complex amplitudes). Because all terms include the factor $e^{j\omega t}$, the general time-dependency is eliminated (see A5).

Substituting

$$v(x,t) \rightarrow \underline{V}(x)\ ;\quad i(x,t) \rightarrow \underline{I}(x)\ ;\quad \partial/\partial t \rightarrow j\omega \quad \text{and} \quad \partial/\partial x \rightarrow d/dx$$

the transmission-line equations in the time-domain become the transmission-line equations for the frequency-domain

$$\frac{d^2\underline{V}}{dx^2} = L'C'(j\omega)^2\,\underline{V} + j\omega\,(R'C'+L'G')\underline{V} + R'G'\underline{V}$$

$$\frac{d^2\underline{I}}{dx^2} = L'C'(j\omega)^2\,\underline{I} + j\omega\,(R'C'+L'G')\underline{I} + R'G'\underline{I} \quad . \qquad (7\text{-}4)$$

Differential equations of this type were encountered for the first time when Oliver Heaviside in 1887 evaluated the signal distortion on telegraph lines. Since then, this differential equation is interdisciplinarily called the *telegraphist's equation* (see Chapter 8). Upon rearranging terms, the transmission-line equations can be written as

$$\frac{d^2\underline{V}}{dx^2} = (R' + j\omega L')(G' + j\omega C')\,\underline{V} \quad \text{or} \quad \frac{d^2\underline{V}}{dx^2} = \underline{\gamma}^2\underline{V}$$

$$\frac{d^2\underline{I}}{dx^2} = (R' + j\omega L')(G' + j\omega C')\,\underline{I} \quad \text{or} \quad \frac{d^2\underline{I}}{dx^2} = \underline{\gamma}^2\underline{I} \quad . \qquad (7\text{-}5)$$

The solution of the transmission-line equations can be obtained using the general exponential solution of d´Alembert´

$$\underline{V}(x) = \underline{A}_1 e^{+\underline{\gamma}x} + \underline{A}_2 e^{-\underline{\gamma}x} \quad . \qquad (7\text{-}6)$$

Hence, we obtain

$$\frac{d^2\underline{V}}{dx^2} = \underline{\gamma}^2\underline{A}_1 e^{+\underline{\gamma}x} + \underline{\gamma}^2\underline{A}_2 e^{-\underline{\gamma}x} = \underline{\gamma}^2\underline{V} \quad . \qquad (7\text{-}7)$$

The complex propagation constant per unit length $\underline{\gamma}$ follows by comparing the coefficients in the boxed equations above as

$$\underline{\gamma} = \sqrt{(R' + j\omega L')(G' + j\omega C')} \qquad (7\text{-}8)$$

$$\underline{\gamma} = \alpha + j\beta$$

Attenuation constant (attenuation per unit length; attenuation: αl) — for α

Phase constant (Phase per unit length; phase: βl) — for β

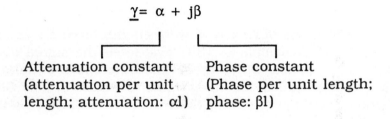

From the aforesaid mesh equation in complex notation,

$$-\frac{d\underline{V}}{dx} = R'\underline{I} + j\omega L'\underline{I} \ , \tag{7-9}$$

together with the general exponential voltage solution $\underline{V}(x)$, we can derive the general current solution

$$\underline{I}(x) = \frac{-\underline{\gamma}}{R' + j\omega L'} \left(\underline{A}_1 e^{+\underline{\gamma}x} - \underline{A}_2 e^{-\underline{\gamma}x}\right) \ . \tag{7-10}$$

The ratio

$$\frac{R' + j\omega L'}{\underline{\gamma}} = \sqrt{\frac{R' + j\omega L'}{G' + j\omega C'}} = \underline{Z}_0 \tag{7-11}$$

is called *characteristic impedance* of the transmission line. The complex coeficients $\underline{A}_1$ and $\underline{A}_2$ are determined from the boundary conditions "voltage and current at the line output":

$$x = l: \quad \underline{V}(x) = \underline{V}_2 \ ; \quad \underline{I}(x) = \underline{I}_2 \ . \tag{7-12}$$

Hence,

$$\underline{V}(x) = \frac{1}{2}(\underline{V}_2 + \underline{I}_2 \underline{Z}_0)\, e^{+\underline{\gamma}(l-x)} + \frac{1}{2}(\underline{V}_2 - \underline{I}_2 \underline{Z}_0)\, e^{-\underline{\gamma}(l-x)} \tag{7-13}$$

$$\underline{I}(x) = \frac{1}{2}(\underline{V}_2/\underline{Z}_0 + \underline{I}_2)\, e^{+\underline{\gamma}(l-x)} - \frac{1}{2}(\underline{V}_2/\underline{Z}_0 - \underline{I}_2)\, e^{-\underline{\gamma}(l-x)} \tag{7-14}$$

and upon multiplication and rearranging

$$\boxed{\begin{aligned} \underline{V}(x) &= \underline{V}_2 \cosh\underline{\gamma}(l-x) + \underline{I}_2 \underline{Z}_0 \sinh\underline{\gamma}(l-x) \\ \underline{I}(x) &= \underline{I}_2 \cosh\underline{\gamma}(l-x) + \underline{V}_2/\underline{Z}_0 \sinh\underline{\gamma}(l-x) \end{aligned}} \ . \tag{7-15}$$

The discussion of the solutions for certain operational modes, for example open and short circuit at the line´s end is found in most texts on applied electromagnetics. It must be pointed out here that this chapter is not supposed to introduce readers to transmission line theory; rather, it presents to them an extremely powerful generic differential equation.

Exploiting the transmission-line concept further, we can find criteria which will allow us to determine whether a problem must be treated as a system with distributed parameters or can be dealt with as a lumped circuit.

The distinction between *electrically long* and *electrically short* lines is possible either in the *frequency-* or *time-domain.*

Definition in frequency-domain (time-harmonic quantities):

Let a sinusoidal voltage with frequency 60 Hz be applied to an infinitely long line. When the source voltage goes through zero a snap-shot will document a voltage distribution as depicted in Figure 7.2.

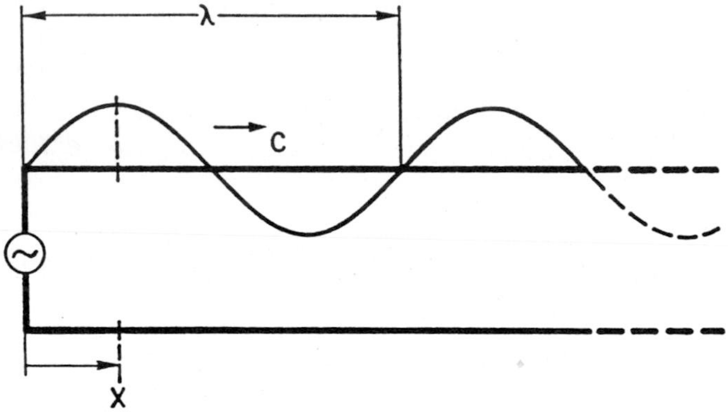

Figure 7.2: Voltage distribution along an electrically long loss-free transmission line.

Depending on the coordinate x, the voltage along the line may possess different values. Changes of the source voltage will be felt at a position x only after traveling a time $t=x/c$ (for $\varepsilon=\varepsilon_0$ and $\mu=\mu_0$).

In the frequency-domain a line is considered *electrically long* if the complex amplitude of the voltage depends on its position on the line

$$\underline{V} = \underline{V}(x) \ . \tag{7-16}$$

It is considered *electrically short* if the complex amplitude of the voltage along the line is approximately constant

$$\underline{V} \approx \text{ const. } , \text{ i.e. } x >> \lambda \tag{7-17}$$

In the radio-frequency-region lines with a length

$$l < \lambda/4 \tag{7-18}$$

are considered electrically short. Power systems are more demanding. For instance, if the voltage difference ΔV is desired to not exceed 0.5 per cent, the line length must not be longer than $\lambda/60$.

With the equation for the wave length

$$\lambda = c \cdot T = c/f \tag{7-19}$$

and the reduced propagation velocity of light in matter

$$c_v = \frac{c}{\sqrt{\varepsilon_r}} \tag{7-20}$$

we obtain for high-voltage transmission lines in air

$$\lambda_T = \frac{c}{f} = \frac{300\,000\ (\mathrm{km/s})}{50/\mathrm{s}} = 6000\mathrm{km} \ , \tag{7-21}$$

hence,

$$l_{max} = 100\mathrm{km} \ . \tag{7-22}$$

For cables with $\varepsilon_r \approx 4$

$$\lambda_C = \frac{c_V}{f} = \frac{300\,000\ (\mathrm{km/s})\ /\ \sqrt{4}}{50/\mathrm{s}} = 3000\mathrm{km} \ , \tag{7-23}$$

hence,

$$l_{max} = 50\mathrm{km} \ . \tag{7-24}$$

Definition in time-domain:

Connecting a dc-voltage source to an infinitely long line generates a traveling wave propagating into the line, Figure 7.3. The risetime T_a of the traveling wave is determined by the nature of the switch, the source impedance and the transmission line's characteristic impedance.

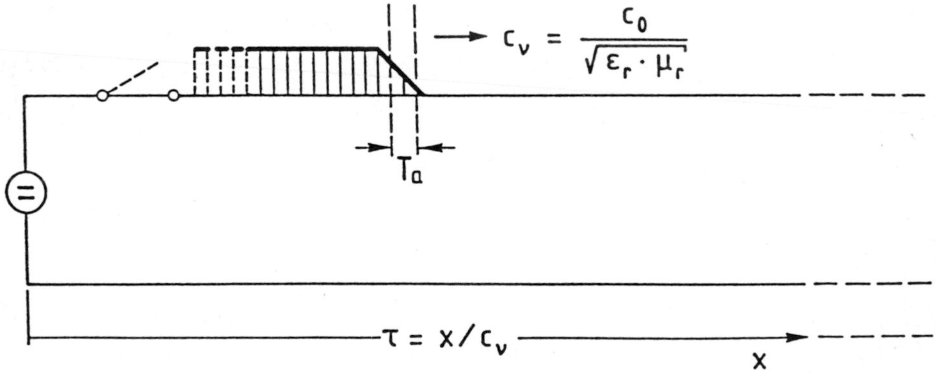

Figure 7.3: Traveling-wave propagation on a long loss-free transmission line; c_V propagation velocity in matter.

In the time domain a line is considered *electrically long* if, at a given instant, the instantaneous value of the voltage is a function of line length,

$$v(t) \Rightarrow v(x,t) \quad . \tag{7-25}$$

In other words, if the traveling wave's risetime is of the same order of magnitude or even shorter than the line's *delay time*. Risetimes $T_a < 10\tau$ require mathematical modeling employing the telegraphist's equation or the wave equation (if loss-free).

A line is considered *electrically short* if the voltage at a given instant possesses the same value everywhere along a line.

$$v(x,t) = v(t) = \text{const}_x \tag{7-26}$$

i.e., if the risetime of the traveling wave is very large compared with the propagation time or delay time

$$T_a >> \tau = \frac{l}{c_v} \quad . \tag{7-27}$$

For risetimes $T_a > 10\tau$, voltage distortions due to the reflections remain below 10 per cent allowing mathematical modeling by ordinary differential equations.

In the time-domain, no distinction between gas-insulated lines and cables is necessary because the different propagation velocities for electromagnetic waves are taken into account by the different propagation times for gas-insulated lines and cables.

8 Typical Differential Equations of Electrodynamics or Mathematical Physics

8.1 Generalized Telegraphist´s Equation

The *telegraphist´s equation* for transmission lines, derived in the previous chapter, represents generically numerous partial differential equations of electrodynamics and many other disciplines, the so-called *differential equations of mathematical physics*. Engineers usually encounter these equations "problem-oriented" and without recognizable links to other familiar equations. The hidden systematics will be disclosed in this chapter by means of some representative examples.

For a generic scalar or vector quantity F, the telegraphist´s equation in the time and frequency domain takes the following form

TIME-DOMAIN – $F(x,t)$	FREQUENCY-DOMAIN – $\underline{F}(x,j\omega)$
$\frac{\partial^2 F}{\partial x^2} = a \frac{\partial^2 F}{\partial t^2} + b \frac{\partial F}{\partial t} + c\,F$	$\frac{\partial^2 \underline{F}}{\partial x^2} = a\,(j\omega)^2 \underline{F} + b\,j\omega \underline{F} + c\,\underline{F}$

.

(8-1)

Its representation in the time-domain allows mathematical modeling of arbitrary time-varying processes F(x,t); its representation in the frequency-domain allows the modeling of sinusoidal time-harmonic states with discrete frequency ω, $\underline{F}(x,j\omega)$. Depending on the physical nature of the quantity F and on the existence of the coefficientes a, b and c, the telegraphist's equation represents a formal mathematical model for a multitude of physical phenomena. Its left side is identical with the one-dimensional Laplace equation; for multi-dimensional problems d^2F/dx^2 is replaced by

$$\nabla^2 F = \frac{d^2F}{dx^2} + \frac{d^2F}{dy^2} + \frac{d^2F}{dz^2} \quad . \tag{8.2}$$

Hence

$$\boxed{\nabla^2 F = a\frac{\partial^2 F}{\partial t^2} + b\frac{\partial F}{\partial t} + c\,F} \quad \boxed{\nabla^2 \underline{F} = a\,(j\omega)^2\underline{F} + b\,j\omega\underline{F} + c\,\underline{F}} \quad . \tag{8-3}$$

8.2 Telegraphist's Equation with a,b>0; c=0

The equations are generally of the type

$$\boxed{\nabla^2 F = a\frac{\partial^2 F}{\partial t^2} + b\frac{\partial F}{\partial t}} \quad \text{or} \quad \boxed{\nabla^2 \underline{F} = a\,(j\omega)^2\underline{F} + b\,j\omega\underline{F}} \quad . \tag{8-4}$$

For instance, substituting for F the fields **E** or **H** and choosing $a=\varepsilon\mu$, and $b=\sigma\mu$, yields the *wave equation for lossy dielectrics* $(\sigma \neq 0)$

$$\nabla^2\mathbf{E} = \varepsilon\mu\frac{\partial^2\mathbf{E}}{\partial t^2} + \sigma\mu\frac{\partial\mathbf{E}}{\partial t} \qquad \nabla^2\underline{\mathbf{E}} = (j\omega)^2\varepsilon\mu\underline{\mathbf{E}} + j\omega\,\sigma\mu\underline{\mathbf{E}} \tag{8-5}$$

or

$$\nabla^2\mathbf{H} = \varepsilon\mu\frac{\partial^2\mathbf{H}}{\partial t^2} + \sigma\mu\frac{\partial\mathbf{H}}{\partial t} \qquad \nabla^2\underline{\mathbf{H}} = (j\omega)^2\varepsilon\mu\underline{\mathbf{H}} + j\omega\,\sigma\mu\underline{\mathbf{H}} \,. \tag{8-6}$$

In ideal dielectrics we have neglected the conduction current with respect to the displacement current. Midway between either typically electric or typically dielectric conductivity, both current-density components need to be considered in Ampere´s circuital law. Then, similarly, as in Sections 6.2 and 6.3, either the electric or magnetic field-strength may be eliminated leading to the equations above.

Substituting for F the potentials **A** or φ, yields the potential equations for the magnetic vector-potential and the electric scalar-potential in conductors and lossy dielectrics

$$\nabla^2\mathbf{A} = \varepsilon\mu\frac{\partial^2\mathbf{A}}{\partial t^2} + \sigma\mu\frac{\partial\mathbf{A}}{dt} \qquad \nabla^2\underline{\mathbf{A}} = (j\omega)^2\varepsilon\mu\,\underline{\mathbf{A}} + j\omega\sigma\mu\,\underline{\mathbf{A}} \tag{8-7}$$

or

$$\nabla^2\varphi = \varepsilon\mu\frac{\partial^2\varphi}{\partial t^2} + \sigma\mu\frac{\partial\varphi}{dt} \qquad \nabla^2\underline{\varphi} = (j\omega)^2\varepsilon\mu\,\underline{\varphi} + j\omega\sigma\mu\underline{\varphi} \tag{8-8}$$

For the explicit derivation of these equations, the reader is referred to the literature.

8.3 Telegraphist´s Equations with a>0, b=0, c=0

The equations are generally of the type

$$\boxed{\nabla^2 F = a \frac{\partial^2 F}{\partial t^2}} \qquad \boxed{\nabla^2 \underline{F} = a\,(j\omega)^2 \underline{F}} \; . \tag{8-9}$$

and generically called the *wave equation* or *vibration equation.* For instance, substituting for F the field strengths **E** or **H** and choosing $a=\varepsilon\mu$ yields the wave equations for electromagnetic waves in loss-free dielectrics ($\sigma=0$), as derived in Section 6.3.1

$$\boxed{\nabla^2 \mathbf{E} = \varepsilon\mu \frac{\partial^2 \mathbf{E}}{\partial t^2}} \qquad \boxed{\nabla^2 \underline{\mathbf{E}} = (j\omega)^2 \varepsilon\mu \underline{\mathbf{E}}} \tag{8-10}$$

$$\boxed{\nabla^2 \mathbf{H} = \varepsilon\mu \frac{\partial^2 \mathbf{H}}{\partial t^2}} \qquad \boxed{\nabla^2 \underline{\mathbf{H}} = (j\omega)^2 \varepsilon\mu \underline{\mathbf{H}}} \; . \tag{8-11}$$

Substituting v and $\underline{V}$, or i and $\underline{I}$, for F yields the wave equation for the propagation of voltage and current traveling-waves along loss-free transmission lines ($R'=0$, $G'=0$, see Chapter 7)

$$\frac{\partial^2 v}{\partial x^2} = L'C' \frac{\partial^2 v}{\partial t^2} \qquad \frac{d^2\underline{V}}{\partial x^2} = (j\omega)^2 L'C'\underline{V} \tag{8-12}$$

$$\frac{\partial^2 i}{\partial x^2} = L'C' \frac{\partial^2 i}{\partial t^2} \qquad \frac{d^2\underline{I}}{\partial x^2} = (j\omega)^2 L'C'\underline{I} \,. \tag{8-13}$$

Substituting for F the displacement (elongation)L or $\underline{L}$ of mass particles and for "a" the reciprocal of the wave propagation velocity v yields the wave equation for mechanical oscillations, e.g. a vibrating string or shock-wave in a water pipe

$$\frac{\partial^2 L}{\partial x^2} = \frac{1}{v} \frac{\partial^2 L}{\partial t^2} \qquad \frac{d^2\underline{L}}{\partial x^2} = \frac{1}{v} (j\omega)^2 \underline{L} \,. \tag{8-14}$$

In contrast to electromagnetic waves, which can propagate also in vacuum, the existence of mechanical waves depends on the existence of matter, in other words, on the existence of mass points possessing the ability to oscillate, or on continuously distributed elastic matter.

8.4 Telegraphist´s Equation with b>0, a=0, c=0

The equations are of the general type

$$\nabla^2 F = b \frac{\partial F}{\partial t} \qquad \nabla^2 \underline{F} = b\, j\omega \underline{F} \tag{8-15}$$

and are generically called the *diffusion equation* or *heat equation*. Note that they also describe skin effects, transport phenomena, and other formally similar effects. For instance, substituting for F the quantities **E**, **H**, or **J** and choosing $b = \sigma\mu$, one obtains the equations for the conduction field in conductors with skin effect derived in Section 6.2.4,

$$\nabla^2 \mathbf{E} = \sigma\mu \frac{\partial \mathbf{E}}{\partial t} \qquad \nabla^2 \underline{\mathbf{E}} = j\omega\sigma\mu\underline{\mathbf{E}} \tag{8-16}$$

or

$$\nabla^2 \mathbf{H} = \sigma\mu \frac{\partial \mathbf{H}}{\partial t} \qquad \nabla^2 \underline{\mathbf{H}} = j\omega\sigma\mu\underline{\mathbf{H}} \tag{8-17}$$

or

$$\nabla^2 \mathbf{J}_C = \sigma\mu \frac{\partial \mathbf{J}_C}{\partial t} \qquad \nabla^2 \underline{\mathbf{J}}_C = j\omega\sigma\mu\underline{\mathbf{J}}_C . \tag{8-18}$$

Substituting the temperature T for F, and $c\rho/\lambda$ for b (with c specific heat, ρ density, and λ thermal conductivity), yields the partial differential equations for space and time-dependent temperature distributions

$$\nabla^2 T = \frac{c\rho}{\lambda}\frac{\partial T}{\partial t} \quad \text{or} \quad \nabla^2 \underline{T} = j\omega \frac{c\rho}{\lambda} \underline{T} . \tag{8-19}$$

Because of this analogy, numerical methods for the calculation of temperature fields and electrostatic potential fields can complement each other, given similar boundary values.

8.5 Helmholtz Equation

Partial differential equations of the aforesaid type are frequently solved by separation of variables, i.e. by a product solution $F(\mathbf{r},t) = u(\mathbf{r})\cdot v(t)$, which decompose the partial differential equation into two similar, ordinary differential equations. Then, in one of these equations only the position occurs as an independent variable, in the other only the time occurs as an independent variable. In other words, space and time dependency become uncoupled. The equation for the space dependency may be brought into the form

$$\boxed{\nabla^2 u + k^2 u = 0} \, , \qquad (8\text{-}20)$$

and is called the *Helmholtz equation.*

If we content ourselves with sinusoidal, i.e. time-harmonic, wave forms, the product solution can be of the type

$$F(\mathbf{r},t) = u(\mathbf{r}) \sin\omega t \quad \text{or} \quad F(\mathbf{r},t) = u(\mathbf{r}) \cos\omega t \; . \qquad (8\text{-}21)$$

Both solutions are combined in complex notation,

$$u(\mathbf{r})\, v(t) \;\rightarrow\; \underline{F}(\mathbf{r})\, e^{j\omega t}\,, \tag{8-22}$$

in which $\underline{F}(\mathbf{r})$ corresponds to the complex amplitude of the familiar phasor notation of ac quantities in antenna and shielding theory or power-system analysis (see A5). The solution of the original equation is then obtained as the real part of the complex solution. One can either substitute the complex general solutions into the original equation, whereby the factors $e^{j\omega t}$ on both sides of the equation cancel, or immediately replace the unknown physical quantity by its complex amplitude, and the differential operators $\partial/\partial t$ by $j\omega$ (operational calculus, see also derivation of the transmission-line equations in Chapter 7). In general, the equation for the space dependency of the complex amplitude $\underline{F}(\mathbf{r})$ takes the following form

$$\boxed{\nabla^2 \underline{F}(\mathbf{r}) = a\,(j\omega)^2\,\underline{F}(\mathbf{r}) + b\, j\omega\, \underline{F}(\mathbf{r}) + c\, \underline{F}(\mathbf{r})}\;, \tag{8-23}$$

and upon factoring out $\underline{F}$ and substituting $k^2 = -\,a\,(j\omega)^2 - jb\omega - c$

$$\boxed{\nabla^2 \underline{F} + k^2 \underline{F} = 0} \qquad \underline{F} = \underline{F}(\mathbf{r})\;. \tag{8-24}$$

Obviously, all differential equations of the preceding chapters can be reduced to a Helmholtz type of equation. The Helmholtz equation can also represent other widely used equations, such as the group diffusion equation for the neutron flux in a reactor

$$\nabla^2\Phi + B^2\Phi = 0 \qquad (\Phi \text{ neutron flux, } B^2 \text{ "Buckling"}), \tag{8-25}$$

or the renowned Schroedinger equation in the next section,

$$\nabla^2\Psi + C\Psi = 0 \qquad (\Psi \text{ Wave function, C energy term}). \tag{8-26}$$

The list could be arbitrarily extended. The difference merely lies in the physical quantity's nature, and the parameter k^2 which, depending on the existence of the coefficients a, b, c may be differently structured and will depend on the problem definition, apart from an arbitrary constant ω. Frequently, only the wave equation in complex form is called the Helmholtz equation, i.e. when b,c = 0.

Differential equations of the Helmholtz type exhibit a peculiarity which will be exemplified for a one-dimensional problem,

$$\frac{d^2u(x)}{dx^2} + k^2\, u(x) = 0 \quad . \tag{8-27}$$

Formally, this ordinary differential equation of second order coincides with the equation of a spring mass oscillator

$$\frac{d^2y(t)}{dt^2} + k^2\, y(t) = 0 \quad , \tag{8-28}$$

if k^2 is interpreted as the ratio of spring constant K to oscillating mass m, that is $k^2 = K/m$. However, whereas the latter equation describes an *initial-value problem*, the Helmholtz equation describes a *boundary-value problem*.

- *Initial-value problems* are characterized by the fact that, from the multitude of possible solutions, only that one is chosen whose value and slope at a given starting-time $t=t_0$ (mostly $t_0=0$) coincides with initial conditions $y(t_0)$, $y'(t_0)$ determined by the specific problem. In other words, the initial conditions determine the integration constant which occurs during integration (solution) of the differential equation.

- *Boundary-value problems* are characterized by the fact, that , from the multitude of possible solutions, only that one is choosen which, at *different* positions e.g. x_1, x_2, takes given values $u_1 = u(x_1)$ and $u_2 = u(x_2)$, determined by the specific problem. Hence, in this case the integration constants are determined by boundary values instead of initial values. Regarding multi-dimensional problems, instead of discrete boundary values, functions $u(\mathbf{r})$ (Dirichlet problem) or $\partial u(\mathbf{r})/\partial n$ (Neumann problem) or linear combinations of both can be used to specify boundary conditions.

Last but not least, combined initial value/boundary-value problems exist, for example partial differential equations with position and time as independent variables.

Whereas with initial-value problems k^2 is uniquely determined by the spring constant K and the oscillating mass m, the factor k^2 of the boundary-value problem is an arbitrary constant ω. Hence, when dealing with practical boundary problems with given boundary conditions, one immediately finds that the Helmholtz equation possesses non-trivial solutions *(eigenfunctions)* only for discrete values of k^2 *(eigenvalues* or *characteristic values)*, for example standing waves of electrically long lines, field modes in wave guides for micro- and optical waves, fundamental and harmonics of guitar strings.

Non-steady states, i.e. space- and time-dependent distributions of arbitrary, nonsinusoidal wave processes (transient waves) are described by superposition of all eigenfunctions. Alternatively, a transient space- and time-dependent distribution, for example voltages and currents on a transmission line, can be evaluated from the superposition of traveling waves with opposite propagation directions (traveling wave theory, theory of transient electromagnetic fields).

Finally it should be emphasized that only solutions of homogeneous Helmholtz equations (right side of equation identically 0, *free oscillations*) are called eigenfunctions. If a *forcing function* exists *(forced oscillations)*, the solutions are also determined by the forcing function. The importance of the Helmholtz equation lies in the fact that it is capable of reducing differential equations of hyperbolic and parabolic types, wave and diffusion equations respectively, to elliptic differential equations, e.g. potential equations, for which numerous solution methods are available.

8.6 Schroedinger Equation

With the lucidity gained for typical equations of electrodynamics it is only a small step to a macroscopic understanding of the *Schroedinger equation*. As is well known, certain physical phenomena, for example interference and bending, call for the interpretation of particles in terms of material waves. For instance, electrons orbiting an atomic nucleus can be assigned a wave function $\psi(\mathbf{r},t)$ which describes the space and time-dependent mass distribution or charge distribution in an atomic hull. This function must be a solution of a wave equation

$$\boxed{\nabla^2\psi = a\frac{\partial^2\psi}{\partial t^2}} \quad \text{or} \quad \boxed{\nabla^2\underline{\psi} = (j\omega)^2 a\underline{\psi}} \quad . \tag{8-29}$$

Substituting

$$a = \frac{2m(W - W_{pot})}{\hbar^2 \omega^2} , \qquad (8\text{-}30)$$

where,

- m electron mass
- W total electron energy ($W_{kin}+W_{pot}$) in the nucleus´ field
- W_{pot} potential energy of an electron in the nucleus´ field
- $\hbar$ Planck´s quantum constant divided by 2π
- ω angular frequency of the material wave,

one obtains an equation in time- and frequency-domain which is already very similar to Schroedinger´s equation,

$$\boxed{\nabla^2\psi = \frac{2m(W - W_{pot})}{\hbar^2 \omega^2} \frac{\partial^2\psi}{\partial t^2}} \quad \text{or} \quad \boxed{\nabla^2\underline{\psi} = -\frac{2m(W - W_{pot})}{\hbar^2} \underline{\psi}} .$$

(8-31)

Transposing all terms of the frequency-domain equation to the left side, we even obtain the familiar form of the Helmholtz equation

$$\boxed{\nabla^2\underline{\psi} + \frac{2m(W - W_{pot})}{\hbar^2} \underline{\psi} = 0} . \qquad (8\text{-}32)$$

By now it is well known that this type of equation possesses nontrivial solutions only for specific eigenvalues (see 8.4), which for the Schroedinger equation correspond to the quantized discrete states of energy of an atom (latent in the energy parameter W).

In physics one replaces the total energy W of a particle by $h \cdot \nu$ or $\hbar\omega$. Then, the equation in time domain takes a different form which includes only the first time-derivative

$$\boxed{\frac{\hbar^2}{2m} \nabla^2 \psi + W_{pot}\, \psi = j\, \hbar \frac{d\psi}{dt}} \quad . \tag{8-33}$$

This is the genuine Schroedinger equation. It can be derived by first substituting $W = \hbar\nu$ in the frequency domain, then multiplying through by $j = \sqrt{-1}$, and eventually returning to the time-domain (considering the convention in physics that $\partial/\partial t \to -j\omega$ instead of $\partial/\partial t \to j\omega$, see A5).

The square of the wave function´s magnitude $|\psi|^2$, is by nature a probability density. It describes the differential probability dP with which an electron can be found within a volume dV = dx dy dz

$$dP = |\underline{\psi}|^2 \, dV \quad . \tag{8-34}$$

Regions in which electrons will be met with high probability, e.g.

$$P = \int |\underline{\psi}|^2 \, dV \geq 0{,}9 \quad , \tag{8-35}$$

are called *orbits*. The product of probability density $|\psi|^2$ and the electron charge "-e" yields the charge density distribution.

The Schroedinger equation is not only a very illustrative example for an eigenvalue problem, but simultaneously demonstrates the basic concept of quantum electronics in which probabilities replace deterministic quantities usually encountered in everyday work and classical physics.

8.7 Lorentz Invariance of Maxwell's Equations

In the preceding chapters we have, with great confidence, thoroughly distinguished between, *electric* and *magnetic* fields. However, it should not remain hidden that, eventually, both fields are merely two different manifestations of one phenomenon of nature found in the surroundings of electric charges. The reader knows that charges are surrounded by electric fields, currents by magnetic fields. Currents, however, are nothing else but electric charges in motion. (The displacement current is called current only because of the fact that

$$\int_S \dot{\mathbf{D}} \cdot d\mathbf{S}$$

has the dimension ampere; being more precise, one ought to speak of the surface integral of a time-varying electric flux-density).

The existence of a magnetic field depends only on whether or not the observer can recognize the motion of a charge (which is affected by a possible motion of the observer relative to the charge). If at rest, in addition to the charge's electric field the observer will sense a magnetic field. If moving along with the charge only an electric field will be sensed. In the surroundings of a conductor, an observer fixed to the conductor will only sense a magnetic field because the electric field of the conduction electrons in motion is precisely compensated by the field of the adjacent positive atomic nuclei of the crystal lattice.

Electromagnetic phenomena manifest themselves in various reference systems with different electric and magnetic field strengths. By means of the *Lorentz transformation* these can be converted into each other. For instance, the following relations exist between field strengths of a fixed coordinate system x y z and a primed coordinate system x'y'z' moving with constant velocity v in x direction (inertial system)

$$E_x = E_x' \qquad E_y = \frac{E_y' + v\,B_z'}{\sqrt{1 - v^2/c^2}} \qquad E_z = \frac{E_z' - v\,B_y'}{\sqrt{1 - v^2/c^2}}$$

$$B_x = B_x' \qquad B_y = \frac{B_y' - E_z'\,v/c^2}{\sqrt{1 - v^2/c^2}} \qquad B_z = \frac{B_z' + E_y'\,v/c^2}{\sqrt{1 - v^2/c^2}}. \tag{8-36}$$

For the charge density ρ' and the current density $\mathbf{J}'$ of a particle beam propagating in the x direction, the following equations apply

$$\rho' = \frac{\rho - J_x\,v/c^2}{\sqrt{1 - v^2/c^2}}$$

$$J_x' = \frac{J_x - v\,\rho}{\sqrt{1 - v^2/c^2}} \qquad J_y' = J_y \qquad J_z' = J_z \tag{8-37}$$

The scalar potential φ and the magnetic vector-potential **A** of a charge propagating in the x direction are given by

$$\varphi' = \frac{\varphi - v A_x}{\sqrt{1 - v^2/c^2}} \qquad - \qquad -$$

$$A_x' = \frac{A_x - \varphi/v}{\sqrt{1 - v^2/c^2}} \qquad A_y' = A_y \qquad A_z' = A_z \qquad (8\text{-}38)$$

Utilizing only quantities of one particular coordinate system, Maxwell's equations hold for any coordinate system without change, i.e. they are invariant with respect to the Lorentz transformation. This is also true for the equation of forces on particles exerted by electric and magnetic fields,

$$\boxed{\mathbf{F} = \mathbf{F}_{el} + \mathbf{F}_{mag} = Q\,\mathbf{E} + Q(\mathbf{v} \times \mathbf{B}) = Q(\mathbf{E} + \mathbf{v} \times \mathbf{B})} \quad . \qquad (8\text{-}39)$$

Depending on the reference system, this force can exclusively be of either electric or magnetic nature.

The invariance of Maxwell's equations becomes clearer yet when we switch to the four-dimensional vector space of the *space-time continuum*. For instance, considering the wave equation of the scalar potential in free space (see 8.1), we obtain with

$$\sigma = 0 \text{ and } c = 1/\sqrt{\varepsilon_0 \mu_0}$$

$$\frac{\partial^2 \varphi}{\partial x^2} + \frac{\partial^2 \varphi}{\partial y^2} + \frac{\partial^2 \varphi}{\partial z^2} = \frac{1}{c^2}\frac{\partial^2 \varphi}{\partial t^2} \quad . \qquad (8\text{-}40)$$

We generalize this equation by substituting new equally entitled coordinates for x,y,z, and t (world coordinates of the *space-time continuum*)

$$x_1 = x\,,\ x_2 = y,\ x_3 = z,\ x_4 = jct\,. \tag{8-41}$$

Then the wave equation becomes the four-dimensional Laplace equation

$$\frac{\partial^2\varphi}{\partial x_1{}^2} + \frac{\partial^2\varphi}{\partial x_2{}^2} + \frac{\partial^2\varphi}{\partial x_3{}^2} + \frac{\partial^2\varphi}{\partial x_4{}^2} = 0 \tag{8-42}$$

or

$$\Box\varphi = 0 \tag{8-43}$$

in which the operator $\Box$ corresponds to the Laplacian $\nabla^2 = \Delta$ generalized to four dimensions.

Power engineers, familiar with computer-aided power system analysis or with the state-variable concept of automatic control theory, will nod approvingly and will not try to visualize the fourth dimension in an elementary geometrical sense as an actually abundant further space coordinate. They are familiar with vectors having 20 or even a thousand components, e.g. in form of the node-voltage vector of a high-voltage transmission grid (voltage profile). They can further visualize that, in a similar way, a four-dimensional Laplace equation for the magnetic vector potential can be defined

$$\boxed{\Box\ \mathbf{A} = 0}\,. \tag{8-44}$$

Further, since electric and magnetic fields have essentially a common origin, it is tempting to combine the magnetic vector potentials' three-component vector A_x, A_y, A_z and the scalar potential φ into a single electromagnetic potential. In a like manner one can combine the three component vectors J_x, J_y, J_z and the charge density ρ into a *single electromagnetic current density*.

Finally, the equations by which fields can be evaluated from their potentials, as well as Maxwell's equations, may be formulated as tensor equations, which are of same form in all inertial systems.

Switching from one reference system to another, of course, the coordinates of a point will change according to the Lorentz transformation. However, the distance between two points defined analogously to the distance in three-dimensional space will not. This distance in four-dimensional space is called *space-time-distance* or *space-time-intervall*, Figure 8.1.

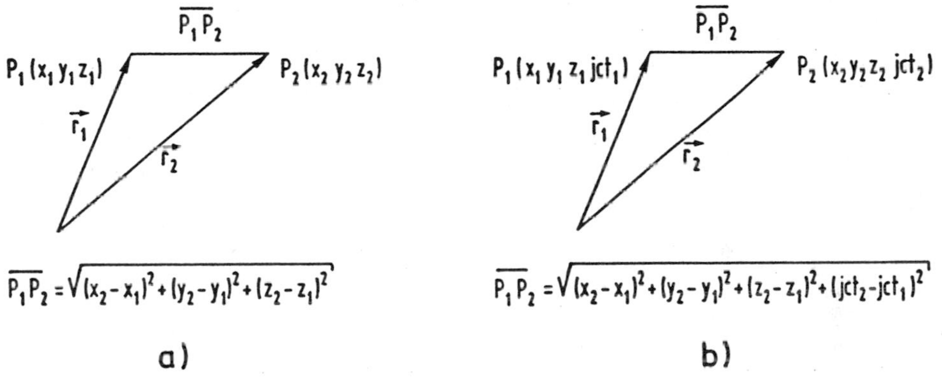

Figure 8.1: a) Definition of a distance in a three-dimensional coordinate system,
b) formal expansion to the generalized distance of four-dimensional space.

If several observers look upon one and the same space-time-distance from different systems, each observer will perceive different contributions of the space and time coordinates. In contrast to classical mechanics, where space and time can be independently transformed, the *Lorentz invariance* of a space-time-distance entails a coupling of space- and time-coordinates.

As a corollary, the appearance of an electromagnetic field depends on an observer's point of view. Neither pure electric and magnetic fields, nor pure charge and current densities exist, which could remain unchanged in different coordinate systems, Nonetheless, for their everyday work the readers may regard electric and magnetic fields as independent phenomena because, for all practical electrical problems, we have $v \ll c$ (exception: relativistic particle beams of accelerators in high-energy physics and pulse-power technology).

Thus, from all that has been said there remains as most important the induction law for moving matter.

$$\oint_C \mathbf{E} \cdot d\mathbf{r} = - \oint_S \frac{\partial \mathbf{B}}{\partial t} \cdot d\mathbf{S} + \oint_C (\mathbf{v} \times \mathbf{B}) \cdot d\mathbf{r} \tag{8-45}$$

Its first term is the voltage induced by a time-varying field in a fixed contour (*transformer induction*); its second term allows the calculation of the voltage induced in a conductor moving in a magnetic field (*motional induction*). The decomposition of the total induced voltage into both components depends on the point of view of the observer.

9 Numerical Calculation of Potential Fields

Potential fields of technical components and apparatus frequently possess only minor symmetry or none at all and do not lend themselves readily to an analytic solution. Computer-aided numerical methods are more powerful, however, their complexity and the multitude of problem-specific algorithms require high initial efforts to get started. This chapter attempts to introduce the reader to the jargon and concepts of numerical field calculations and to provide some decision aids for the choice of a method.

The presently used methods may be classified into four major groups:

Finite-Element Methods
Finite-Difference Methods
Source Simulation Methods
Monte Carlo Methods

The first two may be generically called *differential methods*, the latter two *integral methods*. All four methods will be illustrated by means of electrostatic field examples.

9.1 Finite-Element Method

Regarding finite-element methods, one distinguishes between the *direct method*, *variational method*, and the methods of *weighted residuals* (Galerkins method) and *energy balance*, depending on how an individual element´s characteristics are obtained. For the numerical calculation of potential fields the variational approach is frequently used.

Whereas the ordinary determination of the extreme of a function asks

- for which values x_v does a function f(x) exhibit a maximum or minimum?

variational calculus asks

- for which functions f(x) does a *functional* X(f(x)), i.e. a function of a function, exhibit a maximum or minimum?

In both cases one obtains the solution by differentiating and setting the derivative equal to zero.

Customarily, the functional for the electrostatic field is the potential energy stored in the field. As is well known, charges on electrodes are distributed such that the potential electric energy becomes a minimum value (in much the same way as water flows downhill until a plane water level is established). The energy stored in the volume element ΔV, depicted in Figure 9.1a, is calculated as

$$\boxed{W_{\Delta V} = \frac{1}{2}\,\varepsilon \int_{\Delta V} \mathbf{E}^2(\mathbf{r})\, dV} \,, \qquad (9\text{-}1)$$

where, $\frac{1}{2}\varepsilon\,\mathbf{E}^2(\mathbf{r})$ is the energy density at a point $\mathbf{r}$ in an electric field.

With **E**(**r**) = - grad φ(**r**) = - dφ/d**r** we get

$$W_{\Delta V} = \frac{1}{2}\varepsilon \iint_{\Delta V} \left(\frac{d\varphi}{dr}\right)^2 dV = \frac{1}{2}\varepsilon \iint_{\Delta V} \left(\left(\frac{d\varphi}{dx}\right)^2 + \left(\frac{d\varphi}{dy}\right)^2 + \left(\frac{d\varphi}{dz}\right)^2\right) dV \quad . \tag{9-2}$$

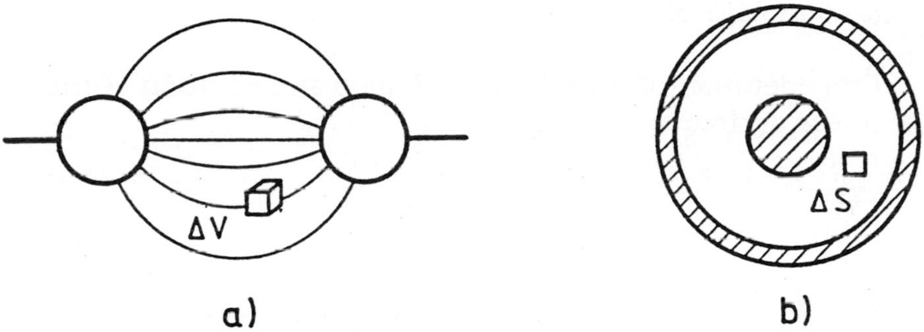

Figure 9.1: Volume elements ΔV and ΔSΔz of an electrostatic field .
a) three-dimensional field,
b) two-dimensional field, z-axis perpendicular to the plane of the paper.

To facilitate our introduction we content ourselves with a two-dimensional field, Figure 9.1b, and obtain the energy per unit length

$$\frac{W_{\Delta V}}{\Delta z} = \frac{1}{2}\varepsilon \int_{\Delta S} \left(\left(\frac{d\varphi}{dx}\right)^2 + \left(\frac{d\varphi}{dy}\right)^2\right) dS \quad . \tag{9-3}$$

$W_{\Delta V}/\Delta z$ is called the functional $X_{\Delta S} = f(\varphi_{\Delta S}(x,y))$ of an element ΔS. The sum of all element functionals, $\Sigma W_{\Delta V}/\Delta z$ is called the functional $X = f(\varphi(x,y)$ of the total field region.

After these preparatory definitions we can focus on the actual finite-element method which proceeds in the following steps:

1. *Discretization* of the field region into individual elements,

2. Development of an *approximation function* for the potential $\varphi_{\Delta S}(x,y)$ within an element,

3. Evaluation of element equations → *element matrices,*

4. Evaluation of the system equation → *system matrix* (assemblage of elements to form a system),

5. Consideration of *boundary conditions* and solution of the linear system of equations.

We will now discuss these steps in detail.

1. Discretization:

The field region is subdivided into n elements of variable size and form, for instance triangles in two-dimensional, or tetrahedrons in three-dimensional fields, Figure 9.2.

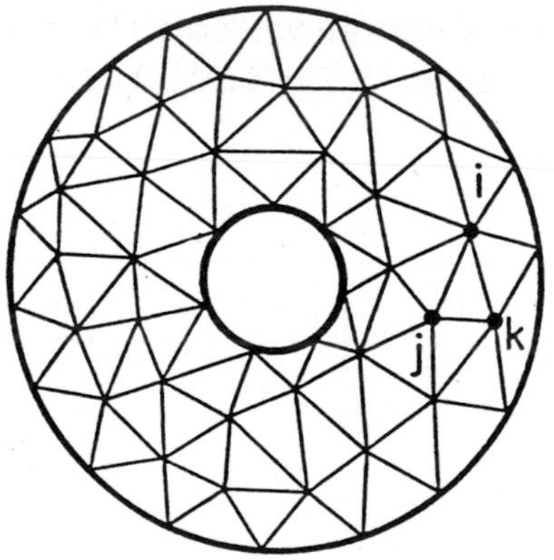

Figure 9.2: Discretization of a two-dimensional field region into triangular elements.

To reduce numerical instabilities, the elements should have as equally long sides as possible. A discretization with curvilinear elements is also possible (e.g. *isoparametric* elements). They allow a smaller number of elements, to model curved boundaries, reducing the input effort considerably. Obviously, the finite-element method is the proper choice for predominantly bounded field regions because the absence of boundaries would require introduction of artificial boundaries (balooning).

2. Approximation functions $\varphi_{\Delta S}(x,y)$ within an element

The simplest approximation of the unknown potential function $\varphi_{\Delta S}(x,y)$ within an element is a linear approach,

$$\varphi_{\Delta S}(x,y) = C_1 + C_2\, x + C_3\, y\,, \tag{9-4}$$

where C_1, C_2, C_3 are a priori unknown coefficients. Higher-order approximation functions are possible or may become necessary, for example for isoparametric, curvilinear elements. For the sake of clarity we restrict ourselves to the linear approximation.

Upon insertion of three coordinate couples $x_i y_i$, $x_j y_j$, $x_k y_k$, the three coefficients C_1, C_2, C_3 can be expressed by the node potentials φ_i, φ_j, φ_k, also referred to as *generalized coordinates*,

$$\left.\begin{aligned} \varphi_i(x_i y_i) &= C_1 + C_2\, x_i + C_3\, y_i \\ \varphi_j(x_j y_j) &= C_1 + C_2\, x_j + C_3\, y_j \\ \varphi_k(x_k y_k) &= C_1 + C_2\, x_k + C_3\, y_k \end{aligned}\right\} \quad \begin{gathered}\text{3 equations for 3 unknowns} \\ C_1\,, C_2\,, C_3\end{gathered} \quad . \tag{9-5}$$

Upon solution of this system of simultaneous equations one obtains the coefficients C_1, C_2, C_3, as functions of the node potentials and node coordinates

$$C_v = f(\varphi_i, \varphi_j, \varphi_k; x_i, x_j, x_k; y_i, y_j, y_k) \quad . \tag{9-6}$$

Substituting the solutions of the linear system of equations for the coefficients of the initial linear approximation function and rearranging terms, we obtain a second version of the approximation function, which approximates the potential inside an element via interpolation through the node potentials $\varphi_i, \varphi_j, \varphi_k$.

$$\boxed{\varphi_{\Delta S}(x,y) = N_i(x,y)\,\varphi_i + N_j(x,y)\,\varphi_j + N_k(x,y)\,\varphi_k} \quad . \tag{9-7}$$

The functions N_i, N_j, N_k are called *interpolation* or *form functions*. They are functions of the nodal coordinates of the respective elements and therefore differ from element to element. For a linear approximation the N_v are calculated from the relation

$$N_i(x,y) = \frac{1}{2\Delta S}\Big((x_j y_k - x_k y_j) + (y_j - y_k)\,x + (x_k - x_j)\,y\Big) \;, \tag{9-8}$$

with cyclically permutated indices in case of $N_j(x,y)$ and $N_k(x,y)$.

Regarding the concept of *generalized coordinates*, it is to be noted that the unique specification of the spatial state of a system with n mass points requires 3n coordinates, one for every *degree of freedom*. Likewise, every nonmechanical system with k degrees of freedom requires k coordinates in order to uniquely specify it. These coordinates need not necessarily be geometrical coordinates, e.g. x,y,z in a Cartesian system, but may manifest themselves as *state variables, node voltages* of an electrical network, or even quantities without any physical meaning. Such non geometric coordinates are called generalized coordinates; their first derivatives are called *generalized velocities*.

3. Evaluation of element equation and element matrices

At this stage the previously mentioned finite-element versions differ in the means by which element equations are derived. For our problem we will employ the variational approach. At first, we take the partial derivative of the approximation function $\varphi_{\Delta S}(x,y)$ with respect to x and y

$$\frac{\partial \varphi_{\Delta S}}{\partial x} = \frac{\partial N_i(x,y)}{\partial x}\varphi_i + \frac{\partial N_j(x,y)}{\partial x}\varphi_j + \frac{\partial N_k(x,y)}{\partial x}\varphi_k \ ,$$

$$\frac{\partial \varphi_{\Delta S}}{\partial y} = \frac{\partial N_i(x,y)}{\partial y}\varphi_i + \frac{\partial N_j(x,y)}{\partial y}\varphi_j + \frac{\partial N_k(x,y)}{\partial y}\varphi_k \ , \qquad (9\text{-}9)$$

and substitute these derivatives into the functional (9-3)

$$\boxed{\frac{W_{\Delta V}}{\Delta z} = X_{\Delta S} = \frac{1}{2}\,\varepsilon \int_{\Delta S} \left(\left(\frac{d\varphi}{dx}\right)^2 + \left(\frac{d\varphi}{dy}\right)^2 \right) dS} \qquad (9\text{-}10)$$

of an element ΔS. In this way we make the element functional a function of the node potentials of an element "ijk"

$$X_{\Delta S} = f(\varphi_i, \varphi_j, \varphi_k) \ . \qquad (9\text{-}11)$$

Just as in ordinary extreme-value calculus, we now differentiate with respect to the potentials and set the derivatives equal to zero

$$\frac{\partial X_{\Delta S}}{\partial \{\varphi\}_{\Delta S}} \triangleq 0, \text{ or in parts, } \frac{\partial X_{\Delta S}}{\partial \varphi_i} \triangleq 0\,; \quad \frac{\partial X_{\Delta S}}{\partial \varphi_j} \triangleq 0\,; \quad \frac{\partial X_{\Delta S}}{\partial \varphi_k} \triangleq 0; \qquad (9\text{-}12)$$

The partial differentiation yields three equations for three unknown potentials φ_i, φ_j, φ_k, hence the following system of linear equations

$$\begin{bmatrix} p_{ii} & p_{ij} & p_{ik} \\ p_{ji} & p_{jj} & p_{jk} \\ p_{ki} & p_{kj} & p_{kk} \end{bmatrix} \begin{Bmatrix} \varphi_i \\ \varphi_j \\ \varphi_k \end{Bmatrix} = 0$$

$$\boxed{[\, p_{\Delta S} \,]\, \{\varphi_{\Delta S}\} = 0} \quad . \qquad (9\text{-}13)$$

The cooefficients of the matrix are functions of the nodal coordinates and the element's permittivity ε_r, i.e. $p_{mn} = F(x_i,x_j,x_k;y_i,y_j,y_k;\varepsilon_r)$. For homogeneous dielectrics ε_r may be factored out.

Because of the physical nature of its coefficients the matrix is called the *permittivity matrix*, or in the three-dimensional case, the *capacitance matrix* (in the latter case, Δz remains on the equation's right side). This matrix is the analogue of the *stiffness matrix* in mechanics. Each element possesses its own permittivity matrix in which each row describes an element node and its relationship to the two other element nodes.

4. Evaluation of the system equation and the system matrix

In this step we will assemble the n element matrices into a system matrix. Recalling that the total field energy must equal the sum of all element energies, one determines the system functional $X=f(\varphi(x,y))$ as the sum of all element functionals $X_{\Delta S} = f(\varphi_{\Delta S}(x,y))$.

$$X = f[\varphi(x,y)] = \Sigma X_{\Delta S} = f(\varphi_{1\,..\,n};\ x_{1\,..\,n};\ y_{1\,..\,n};\ \varepsilon_{r1\,..\,n}) \qquad (9\text{-}14)$$

The partial differentiation of the system functional with respect to all node potentials,

$$\frac{\partial W}{\partial\{\varphi\}} \stackrel{\Delta}{=} 0\,, \qquad (9\text{-}15)$$

that is repeating the minimizing process shown in the preceding section for all elements, yields n equations for n unknown node potentials $\varphi_{1\,...\,n}$, i.e. the linear system of equations

$$\boxed{[P]\ \{\varphi\} = 0} \,. \qquad (9\text{-}16)$$

The coefficients of the system matrix are obtained by summing up all related coefficients of the element matrices. Each row of the system matrix describes an element node and its links to the element nodes of all other adjacent elements

$$P_{mn} = \Sigma\, p_{mn} \quad . \qquad (9\text{-}17)$$

Considering for instance node 10 in Figure 9.3, a diagonal element of the system matrix, e.g $P_{10,10}$, is obtained as follows:

$$P_{10,10} = p_{10,10(1)} + p_{10,10(2)} + p_{10,10(3)} + \ldots + p_{10,10(6)} \,.$$

An off-diagonal element of the system matrix, e.g. $P_{10,12}$, is obtained as

$$P_{10,12} = p_{10,12(2)} + p_{10,12(3)} \quad . \qquad (9\text{-}18)$$

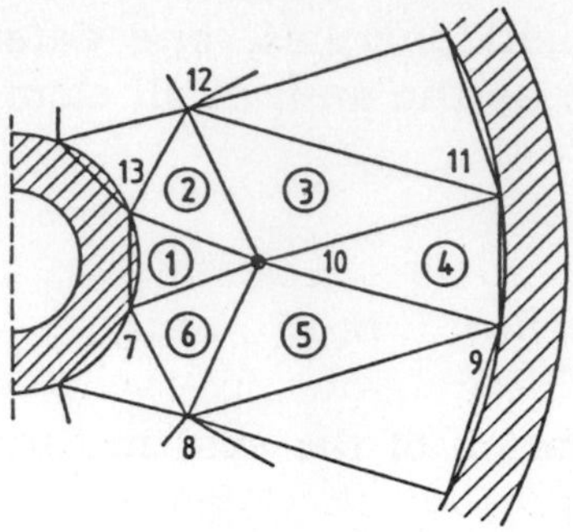

Element ③

$$\begin{bmatrix} P_{10,10} & P_{10,11} & P_{10,12} \\ P_{11,10} & P_{11,11} & P_{11,12} \\ P_{12,10} & P_{12,11} & P_{12,12} \end{bmatrix} \begin{Bmatrix} \varphi_{10} \\ \varphi_{11} \\ \varphi_{12} \end{Bmatrix} = 0$$

Element ②

$$\begin{bmatrix} P_{10,10} & P_{10,12} & P_{10,13} \\ P_{12,10} & P_{12,12} & P_{12,13} \\ P_{13,10} & P_{13,12} & P_{13,13} \end{bmatrix} \begin{Bmatrix} \varphi_{10} \\ \varphi_{12} \\ \varphi_{13} \end{Bmatrix} = 0$$

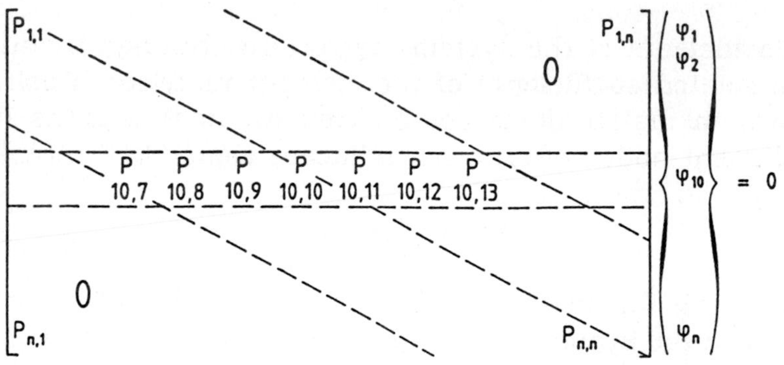

Figure 9.3: Assembling the system matrix from individual element matrices, exemplified for node 10.

Because not every node is connected with all other nodes, the matrix is only sparsely populated, over 99% elements may be zero. By appropriate element numbering it is possible to concentrate most non-zero elements in a narrow band symmetrically distributed with respect to the diagonal (*band matrix*), allowing compact storage.

5. Consideration of boundary conditions

The system matrix P established in the previous paragraph is symmetric and singular, as is the node-admittance matrix Y in circuit analysis, unless a distinction is made between nodes of known and unknown potentials and the equations are rearranged accordingly. In other words, without the introduction of specific boundary conditions, the linear system of equations cannot be uniquely solved. Introducing boundary conditions in numerical solutions of partial differential equations is essentially the same process as determining integration constants or functions in analytical solutions of indefinite integrals or differential equations (see also 8.5). Since we know the potentials of the nodes on boundaries - in the simplest case 0 per cent and 100 per cent - the rows of the boundary nodes can be deleted. However, if we remove all boundary potentials from the nodal potential vector, then all products $P_{mn}\varphi_{boundary}$ containing boundary potentials as factors would also vanish. In order to avoid this, these products are transposed to the equation's right side or we insert a unity row for each boundary potential. In doing so, the square nature of the matrix, lost by deleting rows of boundary nodes, is reestablished resulting in the following linear system of equations

$$[\tilde{P}]\ \{\tilde{\varphi}\} = [P]_{boundary}\ \{\varphi\}_{boundary} \quad , \qquad (9\text{-}19)$$

whose right side is known.

Solving this system of equations for the node-potential vector $\{\tilde{\varphi}\}$ yields the unknown potential values of all element nodes in the field region,

$$\{\tilde{\varphi}\} = [\tilde{P}]^{-1} \, [P]_{boundary} \, \{\varphi\}_{boundary} \quad . \tag{9-20}$$

In practice this linear system of equations is not solved by matrix inversion but by an iterative procedure.

Having evaluated the node potentials, the potential functions $\varphi_{\Delta S}(x,y)$ and the field strength $\mathbf{E}$ = -grad φ in the elements can be determined.

It is to be noted, that a solution of the linear system of equations provides discrete potentials at individual field points. Hence, displaying the results as field plots requires additional programming (post-processing) which connects, for instance, field points with equal potential by interpolation functions, for example spline functions. The latter are piecewise defined functions which, at intersections, possess not only equal values but also equal first derivatives, allowing smooth interpolation.

In case of linear approximation functions, as chosen here, the field strength within an element is constant. Higher-order approximation functions yield field distributions approximating the actual field with better accuracy, Figure 9.4.

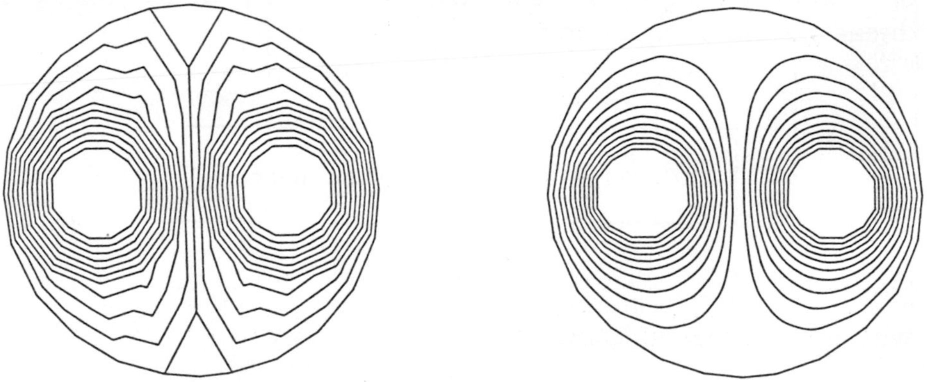

Figure 9.4: Equipotential plot with linear (left) and quadratic approximation function (right).

Finally, it should be mentioned that *multifunctionals* allow simultaneous evaluation of more than one node variable, for example potential *and* field strength.

9.2 Finite-Difference Method

The finite-difference method allows the approximate solution of Laplace's potential equation by degrading its partial derivatives (*quotients of infinitesimals*) to *quotients of incrementals*. It lends itself readily to solutions of two and three-dimensional predominantly bounded field regions. Here, the method is illustrated for two-dimensional electrostatic fields. In the simplest case the field region is discretized by an equidistant point-lattice as in Figure 9.5a

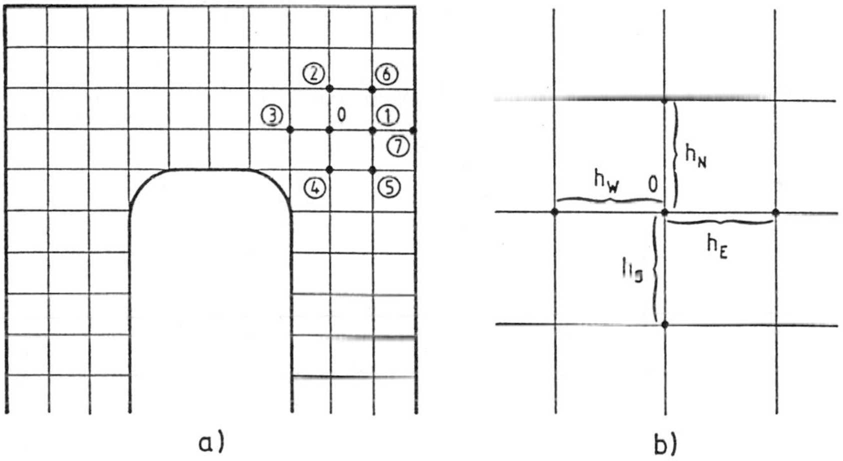

Figure 9.5: Illustration of the finite-difference method
a) discretization of a field region by a point lattice
b) notation of lattice distance for h ≠ const.

Approximating the unknown potential function φ(x,y) in the node "0" by a Taylor series truncated after the second derivative,

$$\varphi(x,y) = \varphi(x_0,y_0) + (x-x_0)\frac{d\varphi}{dx}\bigg|_{x_0y_0} + (y-y_0)\frac{d\varphi}{dy}\bigg|_{x_0y_0} + \frac{(x-x_0)^2}{2}\frac{d^2\varphi}{dx^2}\bigg|_{x_0y_0}$$

$$+ (x-x_0)(y-y_0)\frac{d^2\varphi}{dxdy}\bigg|_{x_0y_0} + \frac{(y-y_0)^2}{2}\frac{d^2\varphi}{dy^2}\bigg|_{x_0y_0} \dots \qquad (9\text{-}21)$$

we obtain the potentials in the adjacent nodes P_1 - P_4 upon insertion of their coordinates

$$\varphi_1 = \varphi(x_0+h\,,y_0) = \varphi_0 + h\frac{d\varphi}{dx} + 0 + \frac{h^2}{2}\frac{d^2\varphi}{dx^2} + 0 + 0$$

$$\varphi_2 = \varphi(x_0\,,y_0+h) = \varphi_0 + 0 + h\frac{d\varphi}{dy} + 0 + 0 + \frac{h^2}{2}\frac{d^2\varphi}{dy^2}$$

$$\varphi_3 = \varphi(x_0-h\,,y_0) = \varphi_0 - h\frac{d\varphi}{dx} + 0 + \frac{h^2}{2}\frac{d^2\varphi}{dx^2} + 0 + 0$$

$$\varphi_4 = \varphi(x_0\,,y_0-h) = \varphi_0 - 0 - h\frac{d\varphi}{dy} + 0 + 0 + \frac{h^2}{2}\frac{d^2\varphi}{dy^2}$$

Adding all four equations yields

$$\varphi_1 + \varphi_2 + \varphi_3 + \varphi_4 = 4\,\varphi_0 + h^2\left(\frac{d^2\varphi}{dx^2} + \frac{d^2\varphi}{dy^2}\right) , \qquad (9\text{-}22)$$

with $h = x - x_0 = y - y_0$.

Because $\Delta\varphi = 0$ (Laplace´s equation), the term in parentheses is identically zero. Hence, the potential in the node "0" can be derived from the four potentials of its adjacent nodes

$$\boxed{\varphi_0 = \frac{1}{4}(\varphi_1 + \varphi_2 + \varphi_3 + \varphi_4)} \quad . \tag{9-23}$$

On electrode boundaries the nodes may not be equidistant, that is h may not be constant. In this case one uses the more general *boundary formula* which can be obtained from the preceding system of equations by substituting the incrementals Δx and Δy for the infinitesimals dx and dy by the actual distances

$$\Delta x = h_W \text{ or } h_E \quad \text{and} \quad \Delta y = h_S \text{ or } h_N \; . \tag{9-24}$$

$$\boxed{\varphi_0 = \frac{\dfrac{\varphi_1}{h_E(h_E+h_W)} + \dfrac{\varphi_2}{h_N(h_N+h_S)} + \dfrac{\varphi_3}{h_W(h_W+h_E)} + \dfrac{\varphi_4}{h_S(h_S+h_N)}}{\dfrac{1}{h_W h_E} + \dfrac{1}{h_N h_S}}} \; . \tag{9-25}$$

For the sake of simplicity, the distances to the adjacent nodes are identified with the directions of the compass. Beginning with the boundary potentials, one can evaluate the potentials of all nodes by *point-wise iteration*.

The equations of all nodes represent a linear system of simultaneous equations, which is conviently obtained by rearranging the boundary formula and applying it to all points of the grid. Upon multiplying the formula by its right side´s denominator and transposing all terms to one side we obtain

$$-\varphi_0\underbrace{\left(\frac{1}{h_W h_E}+\frac{1}{h_N h_S}\right)}_{-h_{00}}+\varphi_1\underbrace{\frac{1}{h_E(h_E+h_W)}}_{h_{0E}}+\varphi_2\underbrace{\frac{1}{h_N(h_N+h_S)}}_{h_{0N}}+\varphi_3\underbrace{\frac{1}{h_W(h_W+h_E)}}_{h_{0W}}+\varphi_4\underbrace{\frac{1}{h_S(h_S+h_N)}}_{h_{0S}}=0$$

or

$$-\varphi_0 h_{00} + \varphi_1 h_{0E} + \varphi_2 h_{0N} + \varphi_3 h_{0W} + \varphi_4 h_{0S} = 0$$

Hence, for all nodes 0 . . . n :

$0: h_{00}\varphi_0 + h_{0E}\varphi_1 + h_{0N}\varphi_2 + h_{0W}\varphi_3 + h_{0W}\varphi_4 \; . \; . \; . \; . \; . \; . \; . \; . \; . = 0$

$1: h_{10}\varphi_0 + h_{11}\varphi_1 \; . \; . \; . \; . \; . \; . \; . \; . \; + h_{15}\varphi_5 + h_{16}\varphi_6 + h_{17}\varphi_7 \; . \; . = 0$

\- - - - - - - - - -

\- - - - - - - - - -

$n: h_{n0}\varphi_0 \; . \; . \; . \; . \; . \; . \; . \; . \; . \; . \; . \; . \; . \; . \; . \; h_{nn}\varphi_n = 0$

In matrix notation

$$\begin{bmatrix} h_{00} & & h_{0n} \\ & \cdot & \\ & & \cdot \\ h_{n0} & & h_{nn} \end{bmatrix} \begin{Bmatrix} \varphi_0 \\ \cdot \\ \cdot \\ \cdot \\ \cdot \\ \varphi_n \end{Bmatrix} = 0$$

$$\boxed{[h]\,\{\varphi\} = 0} \quad . \qquad (9\text{-}26)$$

Because the boundary formula involves only 5 nodes at a time, the matrix is only sparsely populated, for example over 99 per cent of the elements may be zero. Through appropriate node numbering schemes it can be achieved that most non-zero elements are symmetrically distributed in a narrow band around the main diagonal (band matrix) allowing compact storage.

As with the finite-element method, the matrix is singular and the system of equations can be solved only upon introduction of boundary conditions. Again, one deletes the rows of boundary nodes, whose potentials are known, and maintains the square nature of the matrix by transposing the products $h_{mn}\varphi_{boundary}$ to the equation´s right side yielding a linear system of equations with known values on the right side

$$\boxed{[\tilde{h}]\,\{\tilde{\varphi}\} = [h]_{boundary}\,\{\varphi\}_{boundary}} \quad . \qquad (9\text{-}27)$$

Solving this system of equations for the vector $\{\varphi\}$ yields the unknown potentials in the discrete lattice nodes. Potentials between nodes are determined by interpolation.

Expansion to three-dimensional problems is formally simple employing Taylor series of three independent variables leading to the so called *dice formula*

$$\boxed{\varphi_0 = \frac{1}{6}\left[\varphi_1 + \varphi_2 + \varphi_3 + \varphi_4 + \varphi_5 + \varphi_6\right]} \quad . \qquad (9\text{-}28)$$

9.3 Charge (Source)-Simulation Method

In the charge-simulation method a *configuration of simulation charges* is determined whose potential function $\varphi_S(\mathbf{r})$ approximates the true potential function of a physical electrode connected to a voltage source (electrode potential φ_E). In order to find the required configuration, n unknown point charges $Q_1 \ldots Q_n$ are positioned inside the electrode, their positions being defined by the user, Figure 9.6.

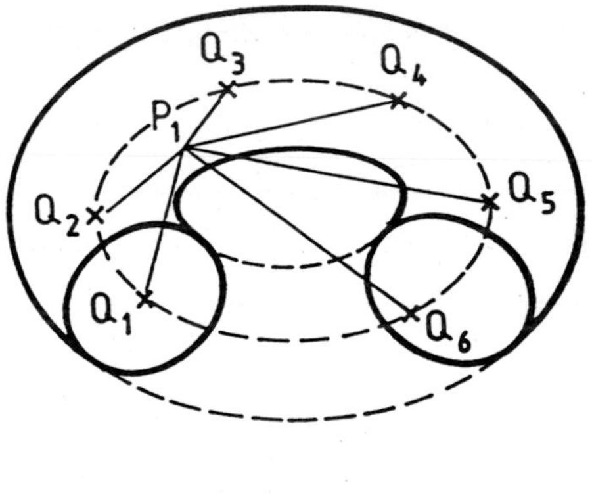

Figure 9.6: Section of a toroidal ring-electrode (e.g. top electrode of a field-controlled impulse voltage divider). Positioning of 6 image charges on the central axis of the toroid.

By means of the superposition principle, one can calculate the potential φ_1 at a point P_1 on the electrode contour (see 4.3),

$$\varphi_1 = \sum_{i=1}^{6} \varphi_{1i} = \frac{Q_1}{4\pi\varepsilon r_{11}} + \frac{Q_2}{4\pi\varepsilon r_{12}} + \ldots + \frac{Q_6}{4\pi\varepsilon r_{16}} \qquad (9\text{-}29)$$

or, with *potential coefficients* (see 4.3)

$$\varphi_1 = \sum_{i=1}^{6} \varphi_{1i} = p_{11} Q_1 + p_{12} Q_2 + \ldots + p_{16} Q_6 \; . \qquad (9\text{-}30)$$

If the six charges ($Q_1 \ldots Q_6$) were known, one could evaluate the potential φ_1 from the geometry. However, in practical electrical problems, we have to cope with the inverse task: given a potential $\varphi_1 = \varphi_E$ in P_1, find 6 image charges $Q_1 \ldots Q_6$ whose superimposed individual potentials coincide with the potential $\varphi_1 = \varphi_E$ in P_1. Since we have only one equation for six unknowns in this case, we reproduce the previous equation for 5 additional contour points $P_2 \ldots P_6$ with identical electrode potential φ_E, resulting in the following linear system of equations

$$p_{11} Q_1 + p_{12} Q_2 + p_{13} Q_3 + p_{14} Q_4 + p_{15} Q_5 + p_{16} Q_6 = \varphi_E$$

$$p_{21} Q_1 + p_{22} Q_2 + p_{23} Q_3 + p_{24} Q_4 + p_{25} Q_5 + p_{26} Q_6 = \varphi_E$$

$$p_{31} Q_1 + p_{32} Q_2 + p_{33} Q_3 + p_{34} Q_4 + p_{35} Q_5 + p_{36} Q_6 = \varphi_E$$

$$p_{41} Q_1 + p_{42} Q_2 + p_{43} Q_3 + p_{44} Q_4 + p_{45} Q_5 + p_{46} Q_6 = \varphi_E$$

$$p_{51} Q_1 + p_{52} Q_2 + p_{53} Q_3 + p_{54} Q_4 + p_{55} Q_5 + p_{56} Q_6 = \varphi_E$$

$$p_{61} Q_1 + p_{62} Q_2 + p_{63} Q_3 + p_{64} Q_4 + p_{65} Q_5 + p_{66} Q_6 = \varphi_E \; .$$

For n chosen image charges one establishes n equations for n arbitrary contour points.

In matrix notation

$$\begin{bmatrix} p_{11} & \cdot & \cdot & \cdot & \cdot & p_{1n} \\ \cdot & & & & & \cdot \\ \cdot & & & & & \cdot \\ \cdot & & & & & \cdot \\ p_{n1} & \cdot & \cdot & \cdot & \cdot & p_{nn} \end{bmatrix} \begin{Bmatrix} Q_1 \\ \cdot \\ \cdot \\ \cdot \\ Q_n \end{Bmatrix} = \begin{Bmatrix} \varphi_E \\ \cdot \\ \cdot \\ \cdot \\ \varphi_E \end{Bmatrix}$$

$$\boxed{[P]\ \{Q\} = \{\varphi_E\}} \quad . \qquad (9\text{-}31)$$

Solving this system of linear, simultaneous equations yields the sought values of charges $Q_1 \ldots\ldots Q_6$

$$\boxed{\{Q\} = [P]^{-1}\ \{\varphi_E\}} \quad . \qquad (9\text{-}32)$$

Thus, we have found the first approximate solution because, with the charges $Q_1 \ldots\ldots Q_6$, the potential $\varphi(\mathbf{r})$ at each point of the field region can be approximately evaluated employing the superposition principle (exact solutions are obtained only for the initially defined contour points of the electrode).

An estimate of the approximate solution´s accuracy is provided by calculating the potentials of other points of the electrode contour, the so-called *check-points*. The potentials calculated for the check-points generally deviate more or less from their nominal value φ_E. Hence, they allow us to judge the quality of the solution resulting from the type amount, and position of the image charges. In a second run, repositioning the point charges, possibly employing more of them, and adding charge types of higher sophistication, e.g. line and ring charges, the approximate solution is upgraded. The number of repeated runs necessarily depends on the expertise of the user.

In order to maintain charge balance, i.e. $\Sigma+Q + \Sigma-Q = 0$, the charge simulation method employs a mirror plane. This results in twice as many potential coefficients. These, however, can be pairwise concentrated in one coefficient, preserving the order of the matrix. The matrix [P] is fully occupied but comparatively small, because its size depends only on the number of image charges.

In contrast to finite-difference and finite-element methods the charge simulation method is particularly well suited for unbounded field regions, for example the spark gap depicted in Figure 9.1 or transmission-line insulators. Two-dimensional problems suggest the use of straight line-charges, three-dimensional problems with rotational symmetry the use of annular line-charges, referred to as ring charges (see 4.3). Employing such charge densities requires integration which can be performed analytically or numerically. Analytical integration allows shorter program execution time. For problems lacking rotational symmetry the ring charges must be segmented or continously varied according to the problem. Multiple dielectrics can be considered by additional charge densities positioned on both sides of a dielectric interface.

A modification of the charge simulation method is the *surface charge method* employing additional *surface-charge densities*, which prove particularly helpful with thin electrodes. Finally, the electrode surfaces can be discretized by plane or curvilinear elements, referred to as the *boundary-element method*. Analytical expression for the surface-charge density is chosen for the elements, which upon analytical or numerical integration allows evaluation of the potentials (see 4.3.2).

9.4 Monte Carlo Method

The Monte Carlo method is based on the *mean-value theorem of potential theory*, in which the potential in the center of a sphere is represented as the mean value of the potentials on its surface, Figure 9.7.

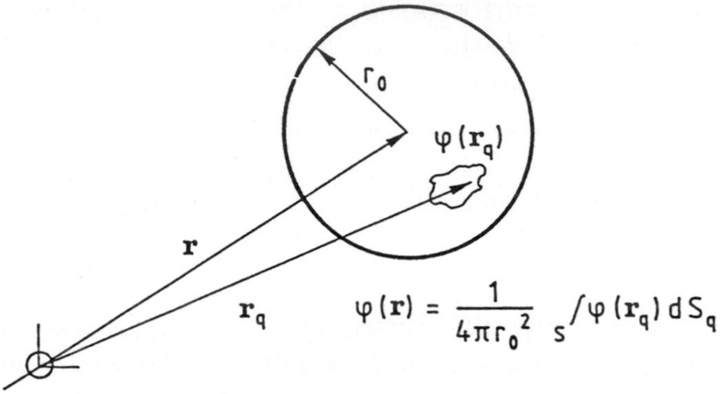

Figure 9.7: Mean-value theorem of potential theory.

Subdividing the surface of the sphere into n surface elements ΔS of equal size, in other words $4\pi r_0{}^2 = n\,\Delta S$, allows approximation of the integral by a sum

$$\varphi(\mathbf{r}) = \frac{1}{n\Delta S}\sum_1^n \varphi_{\Delta S}\ \Delta S\ , \tag{9-33}$$

or

$$\boxed{\varphi(\mathbf{r}) = \frac{1}{n}\sum_1^n \varphi_{\Delta S}}\ . \tag{9-34}$$

The Monte Carlo Method, as applied to numerical field calculation, represents essentially a *statistical interpretation* of this equation. In the simplest case one subdivides the field region by an equidistant point lattice, Figure 9.8.

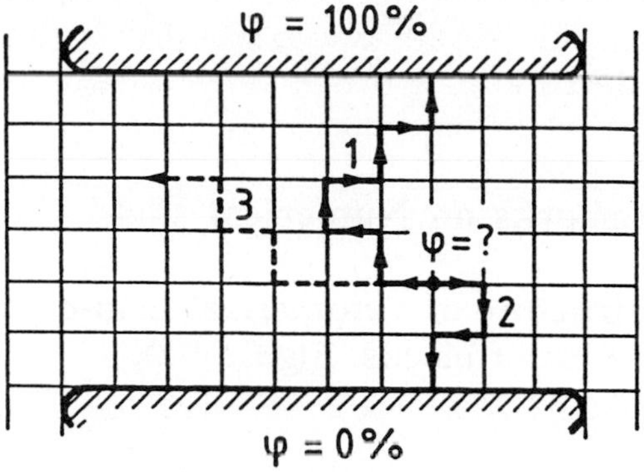

Figure 9.8: Illustration of the Monte Carlo method as used in numerical field calculation. The point grid serves not as discretization of the field region, as is the case with the difference method. It merely provides permissible paths when moving away from the starting point.

Starting from a field point whose potential is sought, one moves to one of four possible adjacent field points, where the direction of propagation in each field point is determined by a random number generator (numbers 1, 2, 3, 4). Performing N runs, one terminates $N_{100\%}$ times at the potential $\varphi_{100\%}$ and $N_{0\%}$ times at the potential $\varphi_{0\%}$. For a reasonable number $N = N_{100\%} + N_{0\%}$ the potential of the starting point is given by

$$\boxed{\varphi(\mathbf{r}) = \frac{1}{N}\left[N_{100\%} \cdot \varphi_{100\%} + N_{0\%} \cdot \varphi_{0\%}\right]} \quad . \tag{9-35}$$

The advantage of the Monte Carlo method lies in its simplicity and the uncomplicated programming technique. However, it demands extensive computer time, because, after N runs, the potential of only a single field point is obtained. An improvement of efficiency is provided by the *floating random-walk* method with varying step width. Compared with the preceding numerical approaches, the Monte Carlo method is understandably less popular.

9.5 General Remarks on Numerical Field Calculation

Basically, the structure of a numerical field-calculation package consists of three major blocks, Figure 9.9.

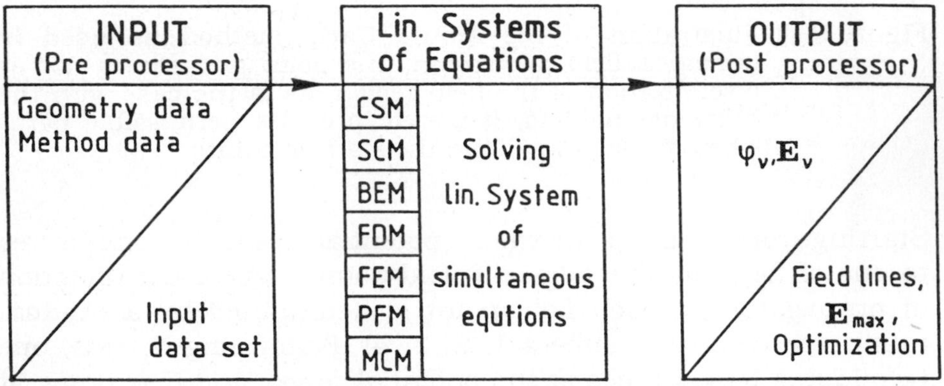

Figure 9.9: Structure of numerical field calculation packages. CSM: Charge-Simulation Method; SCM: Surface-Charge Method; BEM: Boundary-Element Method; FDM: Finite-Difference Method; FEM: Finite-Element Method; PFM: Picture-Frame Method; MCM: Monte Carlo Method.

Depending on the respective method, the blocks may vary in width and certain tasks may be found in either one of adjacent blocks.

- Block 1 establishes an input data set pertaining to the problem geometry, boundary conditions, material properties, and the data of the particular numerical concept.

- Block 2 calculates the matrix coefficients for the selected method and solves the system of simultaneous equations for the vector of unknowns (node potentials, etc).

- Block 3 processes the solution to give field plots of equipotential or field-strength lines, plots of field-strength distributions along specified contours, evaluates contours with constant field-strength, etc.

Only block 2 has been discussed in the preceding sections. If we had to elaborate, plenty of additional details would have to be added. For instance, how the various methods allow consideration of space charges, floating electrodes, different dielectrics and, how storage requirements could be reduced, computing speed increased and numerical accuracy improved, how the accuracy of the approximation could be economically ascertained, when and why numerical instabilities are likely to be encountered, how various approaches could be combined into more powerful tools, etc., etc. An example for a combined approach is the *picture-frame method* (PFM, sometimes called the hybrid method), which exploits the advantages of both the finite-element method and the charge simulation method by using the former within a subregion with arbitrarily selected boundaries, and the latter outside that region.

The pre- and post-processing blocks can be at least as complex and may be the most demanding ones. They determine the degree to which the package is *user friendly* (man machine interaction).

The art of numerical field calculation is still evolving. Future developments are to be expected mainly in pre- and post-processing and will involve *artificial intelligence* techniques.

Appendix

A1 Units and Dimensions

This text uses the rationalized *mksa* system of units which, in its present form, is also referred to as the rationalized *International System of Units* (SI- units). That is, the quantities *length, mass, time,* and *current* are expressed in *meters, kilograms, seconds* and *amperes. Rationalized means that the* factor 4π is not associated with Maxwell's equations but with the potentials, where it can be readily understood when a net flux is divided by the area $4\pi r^2$ of an enclosing sphere (see 4.3). Given the base units above, the *permeability of vacuum (magnetic conductivity)* accounts to

$$\boxed{\mu_o = 4\pi \cdot 10^{-7}\ \mathrm{H/m}} \quad . \tag{A-1}$$

Then, because

$$c = 1/\sqrt{\varepsilon_o\,\mu_o}$$

the *permittivity (dielectric conductivity)* follows as

$$\boxed{\varepsilon_o = 8.85 \cdot 10^{-12}\ \mathrm{F/m}} \quad . \tag{A-2}$$

The *forces* between bodies at rest or in motion are obtained in *newtons; work* and *energy* in *joules*. The units of other electrical and magnetic quantities are collected in table A1.

CONCEPT,	E-FIELD	H-FIELD
Charge	$[Q] = As$ (Coulomb)	----------
Charge density	$[\rho] = \frac{As}{m^3}$	----------
Scalar potential	$[\varphi] = V$	$[\varphi_m] = A$
Vector potential	$[\mathbf{F}] = \frac{As}{m}$	$[\mathbf{A}] = \frac{Vs}{m}$
Voltage	$[V_e] = V$	$[V_m] = A$
Field strength	$[\mathbf{E}] = \frac{V}{m}$	$[\mathbf{H}] = \frac{A}{m}$
Flux	$[\psi] = As$	$[\phi] = Vs$ (Weber)
Flux density	$[\mathbf{D}] = \frac{As}{m^2}$	$[\mathbf{B}] = \frac{Vs}{m^2}$ (Tesla)
Vortex density	$[\mathrm{curl}\ \mathbf{E}] = \frac{V}{m^2}$	$[\mathrm{curl}\ \mathbf{H}] = \frac{A}{m^2}$
Source density	$[\mathrm{div}\ \mathbf{D}] = \frac{As}{m^3}$	$[\mathrm{div}\ \mathbf{B}] = \frac{Vs}{m^3}$
Generalized conductivity	$[\varepsilon] = \frac{[\mathbf{D}]}{[\mathbf{E}]} = \frac{As}{Vm}$	$[\mu] = \frac{[\mathbf{B}]}{[\mathbf{H}]} = \frac{Vs}{Am}$
	e.g. $\varepsilon_0 = 8.8 \cdot 10^{-12} \frac{As}{Vm}$	e.g. $\mu_0 = \frac{4\pi}{10^7} \frac{Vs}{Am}$

Table A1: Units of frequently used electric and magnetic quantities.

Field theory books dicussing Maxwell's equations from a physicist's point of view rather than from an engineering stand point (in other words, ones which deal extensively with electrodynamics of individual charged particles), frequently use the *Gaussian system of units.* This system involves *electrostatic units (esu)* with the prefix "stat-" and *electro-magnetic units (emu)* with the prefix "ab-". On the one hand, this has the advantage of simplifying Coulomb´s law because ε_0 and μ_0 are equal to one, on the other hand, this has the disadvantage that the velocity of light appears in Maxwell´s equations. Today´s electrical engineering science uses the rationalized mksa system exclusively.

A2 Scalar- and Vector Integrals

The reader will recall that a *scalar* or *dot product* of two vectors yields a scalar; a *vector*- or *cross procuct* yields a vector which is perpendicular to the plane defined be the two vectors of the product,

$$\boxed{\mathbf{X} \cdot \mathbf{Y} = X\ Y \cos\alpha = Z} \qquad \boxed{\mathbf{X} \times \mathbf{Y} = \mathbf{Z}} \quad . \tag{A-3}$$

Frequently, the dot in a dot product is not explicitely shown.

Since integrals merely sum up infinitesimal products of the integrand times differential, an integration may yield either a scalar (scalar integral) or a vector (vector integral), depending on the nature of the product.

Scalar integrals: (A-4)

$\int_C \mathbf{X} \cdot \mathbf{dr}$	line integral
$\int_S \mathbf{X} \cdot d\mathbf{S}$	surface integral
$\int_V X \cdot dV$	volume integral

Vector integrals: (A-5)

$\int_C X \cdot \mathbf{dr}$	line integral
$\int_S \mathbf{X} \times d\mathbf{S}$	surface integral
$\int_V (\mathbf{X} \times \mathbf{Y})\, dV$	volume integral

The differential **dr** is to be interpreted in the sense of Figure 3.1. In field theory, occasionally, the differentials d**L** and dL are used instead of $\mathbf{dr}$ and $dr = |\mathbf{dr}|$ (with identical mathematical meaning).

For instance, the scalar integral

$$L = \int_L dL \tag{A-6}$$

yields the true length of the integration path, Figure A1, a. In contrast, the vector integral

$$\mathbf{L} = \int_L d\mathbf{L} \ , \tag{A-7}$$

which sums up all vector differentials $d\mathbf{L}$, yields the distance vector between the boundaries of the integration path, Figure A1 b.

Figure A1: Illustration of the difference between a scalar and a vector line integral.

A3 Vector Operations in Various Coordinate Systems

When deriving Maxwell's equations in differential form, i.e. the equations for source- and vortex-*densities*, we have preferred to write quantities as functions of a position vector **r**, e.g. **D** (**r**), rather than of a triple of three scalar numbers. However, the solution of practical field problems calls for a mathematical model in a particular coordinate system which optimally suits the problem. This system is generally choosen such that its characteristic coordinate surfaces possibly coincide with electrode contours. The most common systems employed are Cartesian (rectangular), cylindrical, and spherical coordinates, Figure A2:

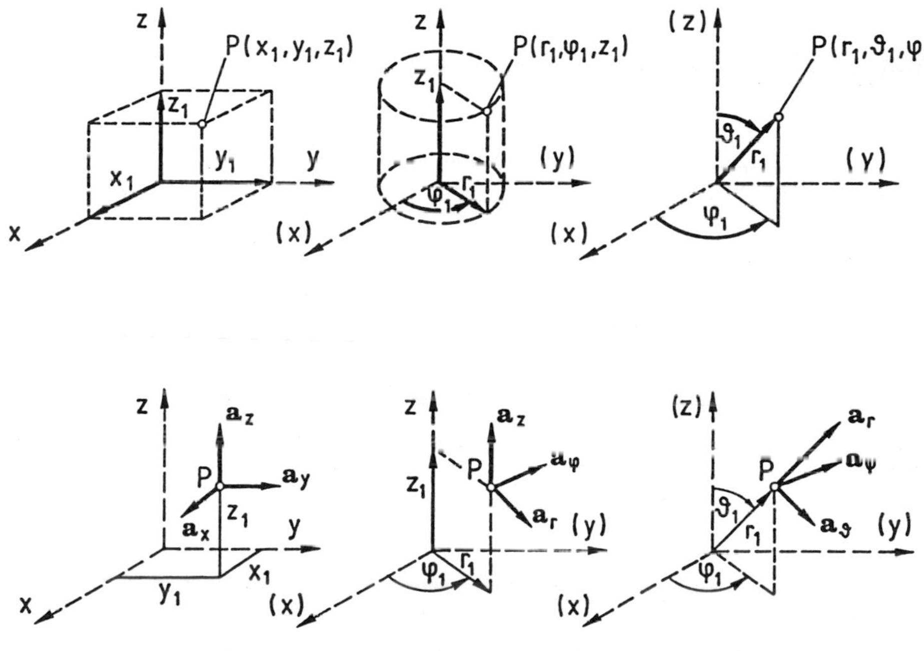

Figure A2: Cartesian, cylindrical, and spherical coordinate systems with respective unit vectors.

Considering the notations in Figure A2, the vortex and source densities, gradient and potential equations of a general vector field $\mathbf{X}(\mathbf{r})$ and a scalar field $U(\mathbf{r})$ take the following forms

Cartesian coordinates x,y z: (A-8)

$$\text{curl}\,\mathbf{X} = \nabla \times \mathbf{X} = \left(\frac{\partial X_z}{\partial y} - \frac{\partial X_y}{\partial z}\right)\mathbf{a}_x + \left(\frac{\partial X_x}{\partial z} - \frac{\partial X_z}{\partial x}\right)\mathbf{a}_y + \left(\frac{\partial X_y}{\partial x} - \frac{\partial X_x}{\partial y}\right)\mathbf{a}_z$$

$$\text{div}\,\mathbf{X} = \nabla \cdot \mathbf{X} = \frac{\partial X_x}{\partial x} + \frac{\partial X_y}{\partial y} + \frac{\partial X_z}{\partial z}$$

$$\text{grad}\,U = \nabla U = \frac{\partial U}{\partial x}\mathbf{a}_x + \frac{\partial U}{\partial y}\mathbf{a}_y + \frac{\partial U}{\partial z}\mathbf{a}_z$$

$$\nabla^2 U = \Delta U = \frac{\partial^2 U}{\partial x^2} + \frac{\partial^2 U}{\partial y^2} + \frac{\partial^2 U}{\partial z^2}$$

Cylindrical coordinates r,φ,z: (A-9)

$x = r\cos\alpha \quad y = r\sin\varphi \quad z = z$

$$\text{curl}\,\mathbf{X} = \nabla \times \mathbf{X} = \left(\frac{1}{r}\frac{\partial X_z}{\partial \varphi} - \frac{\partial X_\varphi}{\partial z}\right)\mathbf{a}_r + \left(\frac{\partial X_r}{\partial z} - \frac{\partial X_z}{\partial r}\right)\mathbf{a}_\varphi + \frac{1}{r}\left(\frac{\partial (rX_\varphi)}{\partial r} - \frac{\partial X_r}{\partial \varphi}\right)\mathbf{a}_z$$

$$\text{div}\,\mathbf{X} = \nabla \cdot \mathbf{X} = \frac{1}{r}\frac{\partial}{\partial r}(rX_r) + \frac{1}{r}\frac{\partial X_\varphi}{\partial \varphi} + \frac{\partial X_z}{\partial z}$$

$$\text{grad}\,U = \nabla U = \frac{\partial U}{\partial r}\mathbf{a}_r + \frac{1}{r}\frac{\partial U}{\partial \varphi}\mathbf{a}_\varphi + \frac{\partial U}{\partial z}\mathbf{a}_z$$

$$\nabla^2 U = \Delta U = \frac{1}{r}\frac{\partial}{\partial r}\left(r\frac{\partial U}{\partial r}\right) + \frac{1}{r^2}\frac{\partial^2 U}{\partial \varphi^2} + \frac{\partial^2 U}{\partial z^2}$$

Spherical coordinates r,ϑ,φ: (A-10)

$x = r \sin\vartheta \cos\varphi \quad y = r \sin\vartheta \sin\varphi \quad z = r \cos\vartheta$

$$\text{curl}\,\mathbf{X} = \nabla\times\mathbf{X} = \frac{1}{r \sin\vartheta}\left(\frac{\partial(X_\varphi \sin\vartheta)}{\partial\vartheta} - \frac{\partial X_\vartheta}{\partial\varphi}\right)\mathbf{a}_r +$$

$$+ \frac{1}{r}\left(\frac{1}{\sin\vartheta}\frac{\partial X_r}{\partial\varphi} - \frac{\partial(rX_\varphi)}{\partial r}\right)\mathbf{a}_\vartheta + \frac{1}{r}\left(\frac{\partial(rX_\vartheta)}{\partial r} - \frac{\partial X_r}{\partial\vartheta}\right)\mathbf{a}_\varphi$$

$$\text{div}\,\mathbf{X} = \nabla\cdot\mathbf{X} = \frac{1}{r^2}\frac{\partial}{\partial r}(r^2 X_r) + \frac{1}{r \sin\vartheta}\frac{\partial}{\partial\vartheta}(X_\vartheta \sin\vartheta) + \frac{1}{r \sin\vartheta}\frac{\partial X_\varphi}{\partial\varphi}$$

$$\text{grad}\,U = \nabla U = \frac{\partial U}{\partial r}\mathbf{a}_r + \frac{1}{r}\frac{\partial U}{\partial\vartheta}\mathbf{a}_\vartheta + \frac{1}{r \sin\vartheta}\frac{\partial U}{\partial\varphi}\mathbf{a}_\varphi$$

$$\nabla^2 U = \Delta U = \frac{1}{r^2}\frac{\partial}{\partial r}\left(r^2\frac{\partial U}{\partial r}\right) + \frac{1}{r^2 \sin\vartheta}\frac{\partial}{\partial\vartheta}\left(\sin\vartheta\frac{\partial U}{\partial\vartheta}\right) + \frac{1}{r^2 \sin^2\vartheta}\frac{\partial^2 U}{\partial\varphi^2}.$$

Wether the result of a differential operation employing the vector ∇ is a scalar or a vector depends on the quality of the operand and on the nature of the procuct (scalar or vector product).

Applying the expanded operators curl, grad, and div to arbitrary scalar and vector fields readily verifies the following vector identities which hold for any coordinate system

$$\text{curl grad}\,U = 0$$

$$\text{div curl}\,\mathbf{X} = 0$$

$$\text{curl curl}\,\mathbf{X} = \text{grad div}\,\mathbf{X} - \nabla^2\mathbf{X}\,. \qquad \text{(A-11)}$$

The first two equations can be readily understood without any mathematical reasoning. In the first case the gradient of a scalar field defines a source field whose vortex density is, by definition, identically zero. In the second case the vortex density of a vector field $\mathbf{x}$ defines another vortex field, whose source density is, by definition, identically zero.

Applying the Laplacian operator on a vector $\mathbf{X}$ in, say, a Cartesian coordinate system, yields the following expansion

$$\nabla^2 \mathbf{X} = \mathbf{a}_x \nabla^2 X_x + \mathbf{a}_y \nabla^2 X_y + \mathbf{a}_z \nabla^2 X_z \ . \qquad \text{(A-12)}$$

Obviously, this result is readily obtained by operating the Laplacian on the scalar vector components X_x, X_y, and X_z. It is to be noted, however, that this is so easy only with Cartesian coordinates. In curvilinear coordinate systems the Laplacian must operate also on the unit vectors.

A4 The Integral Operators $\{curl\}^{-1}$, $\{div\}^{-1}$, $\{grad\}^{-1}$

One basic task in electrodynamics is determining electric and magnetic fields by integrating given vortex- and source-densities, e.g. curl **E** and div **D**. Solving this task is in many respects comparable with the search for an antiderivative

$$F(x) = \int f(x)\,dx + C \ . \qquad \text{(A-13)}$$

Regarding one-dimensional problems, e.g. **E**(x) in a parallel-plate capacitor, there is essentially no difference (see 4.4.1). However, regarding functions of several independent variables, e.g. **E**(x,y,z), integration means solution of partial differential equations which is considerably more difficult.

In sections 4.1, 4.5, 4.6 and 5.3 we could solve partial differential equations in a very elegant way employing innovative inverse integral operators $\{curl\}^{-1}$ and $\{div\}^{-1}$. These operators allowed a very transparent introduction of scalar and vector potentials, which could be grasped immediately.

The new operators have the following meaning:

INTEGRAL OPERATOR $\{curl\}^{-1}$

$$\boxed{\{curl\}^{-1} := \lim_{\Delta r \to 0} \mathbf{n}_r \frac{\Delta \sum_{i=1}^{n} \int_{\Delta S_i} d\mathbf{S}}{\Delta r}} \ , \qquad \text{(A-14)}$$

where $\mathbf{n}_r$ is the unit vector of a contour element $\Delta\mathbf{r} = \mathbf{n}_r \Delta r$ pointing in flow direction.

For example, operating $\{curl\}^{-1}$ on the differential form of Ampere´s law curl **H** = **J**,

$$\{\mathrm{curl}\}^{-1}\ \mathrm{curl}\,\mathbf{H} = \lim_{\Delta r \to 0} \mathbf{n}_r \frac{\Delta \sum_{i=1}^{n} \int_{\Delta S_i} \mathbf{J} \cdot d\mathbf{S}}{\Delta r}$$

and expressing **J** by rot **H** gives

$$\mathbf{H} = \lim_{\Delta r \to 0} \mathbf{n}_r \frac{\Delta \sum_{i=1}^{n} \int_{\Delta S_i} \mathrm{curl}\,\mathbf{H} \cdot d\mathbf{S}}{\Delta r} \quad . \tag{A-15}$$

Stokes´ theorem relates a surface integral to a closed line integral (see 3.6). Thus,

$$\mathbf{H} = \lim_{\Delta r \to 0} \mathbf{n}_r \frac{\Delta \sum_{i=1}^{n} \oint_{C_{\Delta S_i}} \mathbf{H} \cdot d\mathbf{r}}{\Delta r} = \lim_{\Delta r \to 0} \mathbf{n}_r \frac{\Delta \oint_{C} \mathbf{H} \cdot d\mathbf{r}}{\Delta r}$$

$$\boxed{\mathbf{H} = \lim_{\Delta r \to 0} \mathbf{n}_r \frac{\int_{\Delta C} \mathbf{H} \cdot d\mathbf{r}}{\Delta r} = \lim_{\Delta r \to 0} \mathbf{n}_r \frac{\Delta \mathring{V}_m}{\Delta r} = \mathbf{n}_r \frac{dV_m}{dr}} \quad . \tag{A-16}$$

Since a closed line integral does not provide information about the magnitude of the incremental circulation voltage between two distinct points on C the reader may ask how one specifies

$$\Delta \mathring{V}_m = \Delta \oint_C \mathbf{H} \cdot d\mathbf{r} = \int_{\Delta C} \mathbf{H} \cdot d\mathbf{r}$$

Fortunately, this can be readily done by relating $\mathbf{H}(\mathbf{r})$ to the magnetic vector potential $\mathbf{A}(\mathbf{r})$,

$$\mathbf{H}(\mathbf{r}) = \frac{1}{\mu}\mathbf{B}(\mathbf{r}) = \frac{1}{\mu}\,\text{curl}\,\mathbf{A}(\mathbf{r}) \quad ,$$

which can be obtained as a solution of the vector potential equation $\Delta\mathbf{A} = -\mu\mathbf{J}$ (see 5.4),

$$\mathbf{A}(\mathbf{r}) = \Delta^{-1}(-\mu\mathbf{J}(\mathbf{r})) \qquad \text{where} \qquad \Delta^{-1} = \frac{1}{4\pi}\int \frac{\cdots}{|\mathbf{r}-\mathbf{r}_q|}\,dV_q \quad ,$$

yielding

$$\boxed{\mathbf{H}(\mathbf{r}) = \frac{1}{\mu}\,\text{curl}\,\Delta^{-1}\,(-\mu\mathbf{J}(\mathbf{r})) = -\text{curl}\,\Delta^{-1}\mathbf{J}(\mathbf{r})} \quad . \qquad \text{(A-17)}$$

Moreover, from (A-17) we immediately infere the definition of $\{\text{curl}\}^{-1}$ in operator notation,

$$\boxed{\{\text{curl}\}^{-1} := -\,\text{curl}\,\Delta^{-1}} \quad . \qquad \text{(A-18)}$$

For example,

$$\text{curl}\,\mathbf{H} = \mathbf{J}$$

$$\{\text{curl}\}^{-1}\,\text{curl}\mathbf{H} = \mathbf{H} = -\,\text{curl}\,\Delta^{-1}\,\mathbf{J} = \text{curl}\,\mathbf{A}/\mu = \mathbf{B}/\mu \quad .$$

Finally, in order to take care of contour elements with arbitrary direction we multiply (A-16) by $\mathbf{n}_r$

$$\mathbf{n}_r \cdot \mathbf{H} = \frac{dV_m}{dr} ,$$

rewrite the scalar product according to its definition, and rearrange terms. Thus,

$$H = \frac{dV_m}{dr \cos\alpha} ,$$

where α is the angle between **dr** and **H**.

Multipling by $\mathbf{n}_H$ yields

$$\boxed{\mathbf{H} = \frac{dV_m}{dr \cos\alpha} \mathbf{n}_H} \quad . \tag{A-19}$$

INTEGRAL OPERATOR $\{\text{div}\}^{-1}$

$$\boxed{\{\text{div}\}^{-1} := \lim_{\Delta S \to 0} \mathbf{n}_S \frac{\Delta \sum_{i=1}^{n} \int_{\Delta V_i} dV}{\Delta S}} \quad . \tag{A-20}$$

where $\mathbf{n}_S$ is the unit normal vector of an area element $\Delta\mathbf{S} = \mathbf{n}_S \Delta S$ pointing in flow direction.

For example, operating $\{\text{div}\}^{-1}$ on the differential form of Gauss´ law div $\mathbf{D} = \rho$,

$$\{\text{div}\}^{-1} \text{div}\, \mathbf{D} = \lim_{\Delta S \to 0} \mathbf{n}_S \frac{\Delta \sum_{i=1}^{n} \int_{\Delta V_i} \rho dV}{\Delta S}$$

and expressing ρ by div **D** gives

$$\mathbf{D} = \lim_{\Delta S \to 0} \mathbf{n}_S \frac{\Delta \sum_{i=1}^{n} \int_{\Delta V_i} \operatorname{div} \mathbf{D}\, dV}{\Delta S} \quad . \tag{A-21}$$

Gauss´ theorem relates a volume integral to a closed surface integral (see 3.6). Thus,

$$\mathbf{D} = \lim_{\Delta S \to 0} \mathbf{n}_S \frac{\Delta \sum_{i=1}^{n} \oint_{S_{\Delta V_i}} \mathbf{D} \cdot d\mathbf{S}}{\Delta S} = \lim_{\Delta S \to 0} \mathbf{n}_S \frac{\Delta \oint_S \mathbf{D} \cdot d\mathbf{S}}{\Delta S}$$

$$\boxed{\mathbf{D} = \lim_{\Delta S \to 0} \mathbf{n}_S \frac{\int_{\Delta S} \mathbf{D} \cdot d\mathbf{S}}{\Delta S} = \lim_{\Delta S \to 0} \mathbf{n}_S \frac{\Delta \mathring{\Psi}}{\Delta S} = \mathbf{n}_S \frac{d\Psi}{dS}} \quad . \tag{A-22}$$

Since a closed surface integral does not provide information about the magnitude of the partial flux through a distinct surface element ΔS the reader may ask how one specifies

$$\Delta \mathring{\Psi} = \Delta \oint_S \mathbf{D} \cdot d\mathbf{S} = \int_{\Delta S} \mathbf{D} \cdot d\mathbf{S} \quad .$$

Fortunately, this can be readily done by relating **D**(**r**) to the scalar electric potential φ(**r**),

$$\mathbf{D}(\mathbf{r}) = \varepsilon \mathbf{E}(\mathbf{r}) = \varepsilon(-\operatorname{grad} \varphi(\mathbf{r})) \quad ,$$

which can be obtained as a solution of the scalar potential equation $\Delta\varphi = -\rho/\varepsilon$ (see 4.1),

$$\varphi(\mathbf{r}) = \Delta^{-1}(-\rho(\mathbf{r})/\varepsilon) \qquad \text{where} \qquad \Delta^{-1} = \frac{1}{4\pi}\int \frac{\cdots}{|\mathbf{r}-\mathbf{r}_q|}\, dV_q \quad ,$$

yielding

$$\boxed{\mathbf{D}(\mathbf{r}) = -\,\varepsilon\ \text{grad}\ \Delta^{-1}(-\rho(\mathbf{r})/\varepsilon) \;=\; \text{grad}\ \Delta^{-1}\,\rho(\mathbf{r})} \quad . \qquad \text{(A-23)}$$

Moreover, from (A-23) we immediately infere the definition of $\{\text{div}\}^{-1}$ in operator notation

$$\boxed{\{\text{div}\}^{-1} := \text{grad}\ \Delta^{-1}} \quad . \qquad \text{(A-24)}$$

For example,

$$\text{div}\ \mathbf{D} = \rho$$

$$\{\text{div}\}^{-1}\ \text{div}\ \mathbf{D} = \mathbf{D} = \text{grad}\ \Delta^{-1}\,\rho = \text{grad}\ (-\varepsilon\varphi) = -\varepsilon\,\text{grad}\varphi = \varepsilon\,\mathbf{E}.$$

Finally, in order to take care of surface elements with arbitrary orientation we multiply (A-22) by $\mathbf{n}_S$

$$\mathbf{n}_S \cdot \mathbf{D} = \frac{d\Psi}{dS} \quad ,$$

rewrite the scalar product according to its definition and rearrange terms. Thus,

$$D = \frac{d\Psi}{dS \cos\alpha}$$

where α is the angle between **D** and d**S**.

Multiplying by $\mathbf{n}_D$ yields

$$\boxed{\mathbf{D} = \frac{d\Psi}{dS \cos\alpha} \mathbf{n}_D} \quad . \qquad \text{(A-25)}$$

Consequently, we can also define an inverse integral operator $\{grad\}^{-1}$ for the differential operator grad operating on scalar quantities

$$\boxed{\{grad\}^{-1} := \int d\mathbf{r}} \quad . \qquad \text{(A-26)}$$

For example, operating this operator on the relation $\mathbf{E} = -\text{grad}\,\varphi$ between the field strength **E** and the gradient of a scalar potential field $\varphi(\mathbf{r})$,

$$\{grad\}^{-1}\mathbf{E}(\mathbf{r}) = \{grad\}^{-1}(-\text{grad}\varphi(\mathbf{r})) \qquad \text{(A-27)}$$

yields immediately the potential function (s. a. 4.1 and 4.2)

$$-\varphi = \int \mathbf{E} \cdot d\mathbf{r} \ . \qquad \text{(A-28)}$$

The negative sign of φ is due to the fact that we did not include a negative sign in the definition of $\{grad\}^{-1}$ (wich would be appropriate if the operator would be exclusively used in electromagnetics, see 4.1).

Finally, we shall discuss the nature of the solutions obtained via the new operators. As is well kown, differential equations possess an infinite number of solutions consisting of a *general solution* and, regarding one-dimensional problems, an *integration constant*, regarding multi-dimensional problems, additional *integration functions*. The new operators yield the *general solution* of the respective differential equation. Such solutions become a *special solution* only upon taking into account problem-specific boundary conditions or *eo ipso* given assumptions specifying integration constants or integration functions.

Because of the multivaluedness of an integration, pure mathematicians would not consider $\{curl\}^{-1}$, $\{div\}^{-1}$, $\{grad\}^{-1}$ as operators, rather they would call them *canonical inverses*. Today the term operator is reserved for unique functions, e.g. the differential operators curl, div, and grad. However, practical problems are only considered solved when boundary conditions have been taken into account, which make the solution unique. This uniqueness sanctifies the generic concept of integral operators for daily engineering.

A5 Complex Notation for Time-Harmonic Quantities

Mathematical treatment of circuit and continuum problems, e.g. power system analysis, transmission lines, electromagnetic shields etc., is considerably facilitated if we restrict ourselves to sinusoidal, time-harmonic waveforms. Nonsinusoidal transitions from one harmonic state to another or switching transients can be eventually dealt with by applying Fourier transforms to the results found for time-harmonic quantities. Further, when solving partial differential equations by separation of variables, it is frequently convenient to choose a sinusoidal function for the solution´s time component. Once this decision has been made, one complements the sinusoidal function via Euler´s formula to give a complex quantity. This pays off because manipulating complex quantities is much simpler than dealing with theorems governing addition, multiplication etc. of sinusoidal quantities. Eventually, the sought solution is given by the real part of the complex solution.

In detail we proceed along the following steps (see 7):

1. Assume the time-domain solution to be sinusoidal, e.g.

$$v(t) = \hat{v} \cos (\omega t + \varphi_n) \tag{A-25}$$

2. Complement the solution to give a complex quantity

$$\underline{v}(t) = \hat{v} \left[\cos (\omega t + \varphi_n) + j \sin (\omega t + \varphi_n)\right] \tag{A-26}$$

3. Apply Euler´s formula

$$\underline{v}(t) = \hat{v}\, e^{j(\omega t + \varphi_n)} = \underbrace{\hat{v}\, e^{j\varphi_n}}_{\text{complex amplitude (phasor)}} \cdot \underbrace{e^{j\omega t}}_{\text{time factor}} = \underline{V}\, e^{j\omega t} \tag{A-27}$$

complex amplitude (phasor) — $\underline{V}$; *time factor* — $e^{j\omega t}$

4. Substitute for the real physical quantity v(t) the complex quantity $\underline{v}(t)$. While doing this, the factor $e^{j\omega t}$ cancels out; further, differential and integral operations change to multiplications or divisions by the factor $j\omega$.

5. Solve the equation for the complex amplitude $\underline{V}$, then take its real part.

This process eliminates the general time-dependency from partial differential equations, reduces partial differential equations to ordinary differential equations, and changes the latter to algebraic equations. The reader will recognize the similarity with *Laplace transforms.*

Representation of sinusoidal quantities by phasors is an essentail tool in field theory, for instance, when dealing with shielding efficiencies in electromagnetic compatibility (EMC), skin effects, wave guides, electrical machines etc. It should be pointed out that in physics, the sign in the time factor´s exponent is customarily chosen negative,

$$\underline{v}(t) = \underline{V}\, e^{-j\omega t} \quad , \qquad \text{(A-28)}$$

resulting in opposite signs in the frequency-domain (see 8.5). Of course, this does not affect the real solution in the time domain.

Literature

Books (selection):

ALONSO, M und FINN, E.Z., Physik, Amsterdam, Inter European Editions 1977

BOHN, E.K. , Introduction of Electromagnetic Waves, Reading Mass. (USA), Addison Wesley 1968

CHEN, H.C. , Theory of Electromagnetic Waves, New York McGraw-Hill 1983

CHENG, D.K. , Field and Wave Electromagnetics, Reading Mass (USA), Addison-Wesley 1983

CLAYTON, R. P. and NASER, S.A. ,Introduction to Electromagnetic Fields, New York, McGraw-Hill 1982

CRAWFORD, F.S. , Schwingungen und Wellen (Berkeley Physic Kurs Bd. 2 und 3), Braunschweig, Vieweg 1974

EGGES, L, The Classical Electromagnetic Field, Dover Publishers New York, 1972

EINSTEIN, A., Grundzüge der Relativitätstheorie (Nachdruck), Braunschweig/Wiesbaden Vieweg 1984

FEYNMANN, R.P. , Lectures on Physics, Vol.2, Reading Mass. Addison Wesly 1979

HAYT, W.H. , Engineering Electromagnetics, McGraw-Hill Book Comp. 4th edition 1981

JOHNK, C.T.A. , Engineering Electromagnetic Fields and Waves, New York, J. Wiley 1975

KING, R.W.P. and PRUSAD, SH. , Fundamental Electromagnetic Theory and Applications

KOSHLYAKOV, N.S., SMIRNOW, M.M., GLINER, E.B. , Differential Equations of Mathematical Physics, Amsterdam, North-Holland Publ. 1964

KRAUS, J. and CARVER, K.R., Electromagnetics, New York, McGraw-Hill, 3rd edition 1984

LANDAU, L.D. and LIFSHITZ, E.M., The Classical Theory of Fields, Oxford Pergamon Press 1980

MARION J.B. and HEALD M.A., Classical Electromagnetic Radiation, London, Academic Press, 2nd Ed. 1980

NARAYANA RAO, N., Basic Electromagnetics with Applications, Englewood Cliffs, N.J. , Prentice-Hall, 1972

NEFF, H., Basic Electromagnetic Fields, New York, Harper and Row 1981

PARTON, J.E., OWEN, S.J. and RAVEN, M.S., Applied Electromagnetics, Springer Publishers New York, Second edition 1986

PLONUS, M.A., Applied Electromagnetics, New York McGraw-Hill, 1978

SEELY, S., Electromagnetics: Classical and Modern Theory and Applications, New York, M. Dekker, 1979

SKITEK, G.G. and MARSHALL, S.V., Electromagnetic Concepts and Applications, Prentice Hall, Englewood Cliffs N.J. 1982

SOMMERFELD, A., Elektrodynamik (Nachdruck), Thun/Frankfurt Verlag Harri Deutsch 1977

STINSON, D.C., Intermediate Mathematics of Electromagnetics, Englewood Cliffs, N.J., Prentice Hall, 1976

THOMAS, D.T., Egineering electromagnetics, New York, Pergamon Press, 1972

Special Literature on Numerical Calculation of Fields (selection)

BATHE, K.J., Finite Elemente Procedures in Engineering Analysis, by Prentice-Hall, Inc., Englewood Cliffs, New Jersy 1982

CHARI, M.V.K.U., Silvester, G., Finite Elements in Electrical and Magnetic Field Problems, Chichester, New York, 1980

MITRA, R., Numerical and Asymptotic Techniques in Electromagnetics, Springer Publishers Berlin, New York, 1975

PICHLES, I.H., Monte Carlo Field Calculations, Proc. IEE Vol.124 (1977), S.1271-1276

SINGER, H., STEINBIGLER, H. and WEIß, P., Charge Simulation Method for the Calculation of High-Voltage Fields, IEEE Trans. Power Appar. Syst.93 (1974) 1660-1668

STEINBIGLER, H., Anfangsfeldstärken und Ausnutzungsfaktoren rotationssymmetrischer Elektrodenanordnungen in Luft, Dissertation TU München

WEBER, E., Electromagnetic Theory, Dover New York, 1965

WEXLER, A., Some Recent Developments in Field Calculations, IEEE Trans. on Magnetics Vol Mag - 15 (1979), S.1659-1664

Index

Action at a distance, 18
Ampere´s circuital law, 23
  in differential form, 42
Approximation function, 171,173
Attenuation, 144

Boundary conditions, 159,179
Boundary element method, 192
Boundary formula, 183
Boundary-value problem, 159

Capacitance, 3
  matrix, 176
Capacitive fields, 117
Characteristic impedance, 145
Characteristic values, 159
Charge simulation method, 186
Circulation voltage, 24,28
Classification of electric and magnetic vector potentials, 84,94
Complex notation, 211
Complex quantities, 211
Conductance, 3,4
Conduction, 4
  current, 4
  field 4,111
  flux density, 12,16
Conservative fields, 15
Constitutive relations, 10
Continuity Law
  integral form 32
  differential form 48
Coordinate systems,
  cartesian, 199
  cylindrical, 200
  generalized, 173
  rectangular, 199
  spherical, 199
Coulomb gauge, 97
Cross product, 197
Curl, 42

Del operator, 63,200
Dice formula, 186
Differential equations, 150
  of mathematical physics, 150
Differential operator, 199
Diffusion equation, 155
Dirichlet problem, 159
Discretization of the field region, 172,181
Displacement current, 27
Displacement density, 5
Divergence:
  of electric fields, 44
  of magnetic fields, 46
  of conduction fields, 48
Dot product, 195

Eddy currents, 23
Eigenfunctions, 159
Eigenvalues 159
Electric:
  current density, 9,14,111
  field strength, 4,14
  source fields, 15
  vector potential, 84
  voltage, 3,14,67
  vortex fields, 16
Electromagnetic:
  fields, 107
  waves, 125
Electromotive force, 21
Electrostatic fields, 107
Element functionals, 172
Element matrix, 178
Energy densities, 137

Faraday´s induction law, 20
  in differential form, 38
Far-action theory, 18
Field´s work, 65
Finite-element method, 169
Finite-difference method, 181
Floating random-walk method, 192
Flux:
  density, 5
  tubes, 10
  of conduction, 5
Form function, 174

Functional 170

Gauge function, 86,132
Gauss's law,
of the electric field, 30
of the magnetic field, 31
of the conduction field, 32
Gradient, 62
of a scalar field,

Harmonic variations, 211
Heat equation, 155
Helmholtz equation, 156
Hertz potentials, 134
High-impedance fields, 117
High-voltage potential, 67

Inductance, 4
Induction, 55
effect,
Inductive field, 119
Inductive voltage drop, 55
Inertial system, 164
Initial-value problem, 159
Interpolation function, 174
Irrotational fields, 15
Isoparametric elements, 173

Kirchhoff's voltage law, 56

Laplace:
equations, 78
operator, 78
Lamellar fields, 15
Lenz's law, 22
Line charge, 72
Line integral, 197,198
Lorentz:
condition, 132
gauge, 132
invariance, 163
transformation, 132
Low-impedance field, 119

Magnetic:
circulation voltage, 24
field strength, 4,14
flux density, 8
induction, 8,10
scalar potential, 90
vector potential, 94
voltage, 4,14,24
vortex fields, 16
Magnetomotive force, 24
Magnetostatic fields, 109
Maxwell's equations,
in complex notation, 52
in differential form (point form), 37
in integral form, 19
Mean-value theorem, 190

Monte Carlo method, 190
Motional induction, 168
Multifunctionals, 181

Nabla, 63,200
Near-action theory, 18
Net flux through a closed surface, 30
Neumann problem 159
Newton potentials, 16
Numerical field calculation, 169
Node potentials, 172
Nonconservative fields, 15
Nonstationary fields, 107,125

Penetration depth, 124
Permeability of vacuum, 194
Permeance, 4
Permittivity of vacuum 194
matrix, 176
Phase constant, 144
Picture-frame method, 193
Point-wise iteration, 183
Potential:
coefficient, 71,187
concept, 65
equation, 77
Potential function,
of a general charge, 69
of a line charge, 72
Power-flux density, 138
Poynting vector, 138
Propagation time, 130,149
Propagation velocity, 125,147

Quasi-static:
conduction fields,
electric fields, 115
magnetic fields, 118
problems, 104,115
Quasi-stationary fields, 106,115

Rationalized international system of units, 103
Retarded potentials, 129
Rotational fields, 15

Scalar fields, 1
Scalar product, 5,197
Schroedinger equation, 160
Self-induced voltages, 55
Separation of variables, 156
Sinusoidal wave forms, 211
Skin depth, 124
Skin effect, 121
Source-density,
of conduction fields, 48
of electric fields, 44
of magnetic fields, 46
Source-strength,

of conduction fields, 32
of electric fields, 30
of magnetic fields, 31
Space-time distance, 167
Space-time continuum, 165
Space-time interval, 167
Static conduction fields (dc current-conduction field), 111
Stationary fields, 107
Stokes´ theorem, 53
Superposition principle, 72,76
Surface charge:
densities, 75,189
method, 189
Surface integral, 8,197
System of linear equations, 177,179,184
System matrix, 172,178

Telegraphist´s equation, 139,150
Time-harmonic quantities, 211
Transformer induction, 168
Transient electromagnetic fields, 118,125
Transmission-line equations, 139
Traveling wave, 148

Uniqueness, 6,68,96

Variational calculus, 170
Variational method, 170
Vector fields, 15
Vibration equation, 153
Volume integral, 197
Vortex-density,
of electric vortex fields, 38
of magnetic vortex fields, 42
Vortex field, 16
Vortex strength,
of electric vortex fields, 20
of magnetic vortex fields, 23
Vortices, 17

Wave:
equation, 125
function, 160